Problem Solving Workbook

with Reading Strategies

TEACHER'S EDITION
Grade 1

Harcourt Brace & Company

Orlando • Atlanta • Austin • Boston • San Francisco • Chicago • Dallas • New York • Toronto • London

http://www.hbschool.com

CONTENTS

One-to-One Correspondence

Check children's drawings.

1.

2.

3.

4.

Choose the best hat.
Draw one hat for each person.

Harcourt Brace School Publishers

More and Fewer

Check children's drawings.

I.

Answers will vary. Should show 5 or more windows.

2.

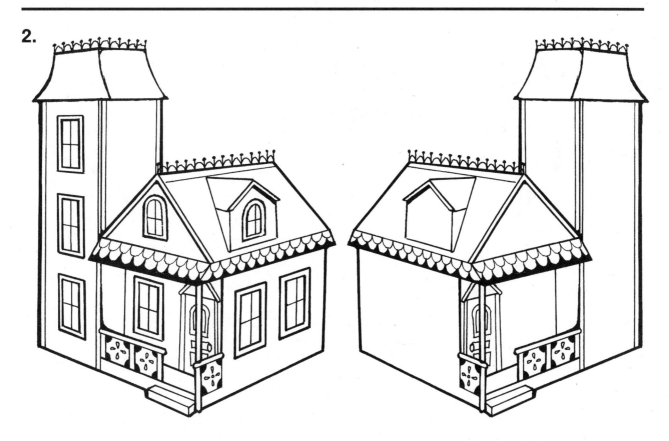

Answers will vary. Should show 7 or fewer windows.

Draw more windows on the house.
Draw fewer windows on the house.

Numbers Through 5

2

4

5

3

1

Count how many objects
each person is juggling.
Write the number of objects.

PS6   PROBLEM SOLVING

Harcourt Brace School Publishers

Name ____________________

Numbers Through 9

7 

9 

8 

6 

Count how many of each.
Write the number.

Ten

Guess   Check

_____ 9

Guess   Check

_____ 10

Guess   Check

_____ 10

Guess   Check

_____ 10

Guess how many of each.
Then count to check your answer.

Greater Than

		Inside	Outside

1.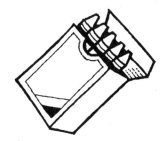

2.  

3.

4.

5.

Write how many are inside and how many are outside.
Circle the number that is greater.

Less Than

red

orange

green

Color orange the group of vegetables that has 1 less than the group of pumpkins.
Color red the group of vegetables that has 1 less than the group of peppers.
Color green the group of vegetables that has 1 less than the group of tomatoes.

PS10   PROBLEM SOLVING

Order Through 10

Number the stepping stones in order.
Color the shortest path blue.
Color the longest path red.

PROBLEM SOLVING   PS11

Ordinal Numbers

Start

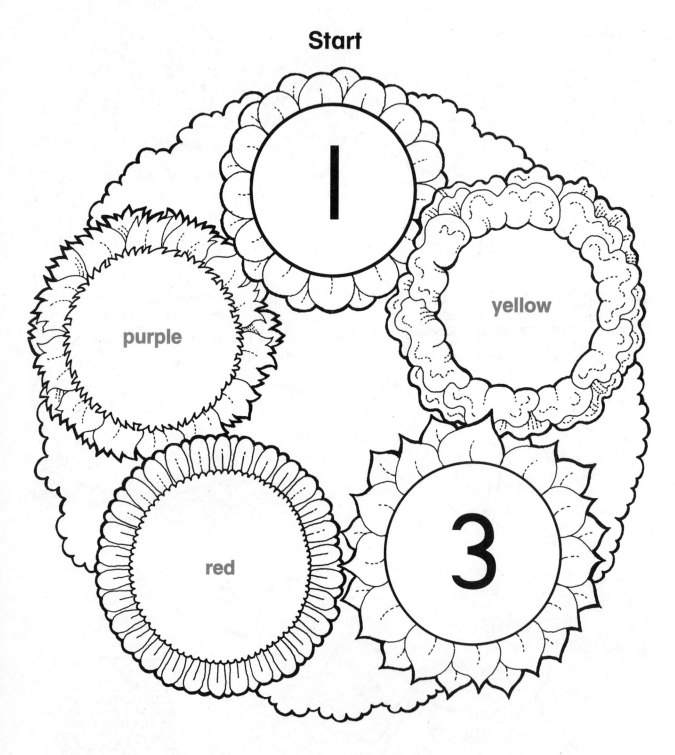

Go around the wreath. Color the second flower yellow.
Color the fourth flower red.
Color the fifth flower purple

PS12   PROBLEM SOLVING

Modeling Addition Story Problems

Draw and . **Check children's work.**
Write how many in all.

1. Show one way to make 3.

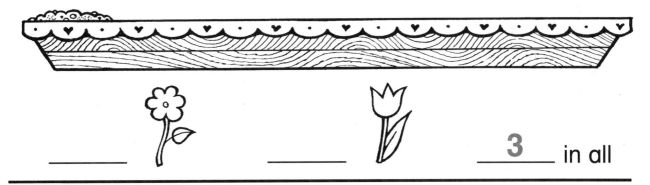

_____   _____   _3_ in all

2. Show one way to make 6.

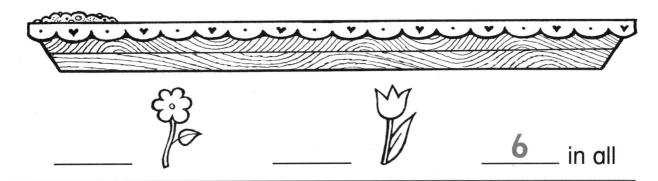

_____   _____   _6_ in all

Make up an addition story.
How many in all? Mark your answer.

3.

| 1 | 2 | 3 |
| ○ | ⬤ | ○ |

4.

| 4 | 5 | 6 |
| ○ | ⬤ | ○ |

Adding 1

Draw 1 more.
Write the sum. **Check children's drawings.**

1. 1 fish.
   1 more comes.

$$1 + 1 = \underline{\quad 2 \quad} \text{ fish}$$

2. 3 cats.
   1 more comes.

$$3 + 1 = \underline{\quad 4 \quad} \text{ cats}$$

Choose the correct answer.

3. 5 ducks.
   1 more comes.
   How many ducks?

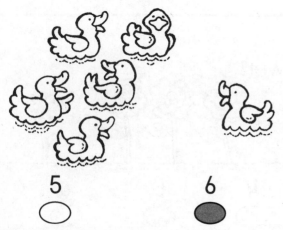

5	6
◯	⬤

4. 4 sheep.
   1 more comes.
   How many sheep?

4	5
◯	⬤

Adding 2

Draw 2 more.
Write the sum. **Check children's drawings.**

1. 1 cow.
   2 more come.

$$1 + 2 = \underline{}3 \text{ cows}$$

2. 2 bears.
   2 more come.

$$2 + 2 = \underline{}4 \text{ bears}$$

Choose the correct answer.

3. Which hat has 2 more than 1?

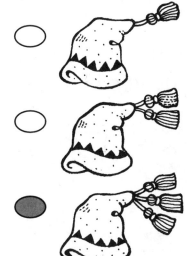

4. Which bunch has 2 more than 3?

PROBLEM SOLVING   PS15

Name _____

Using Pictures to Add

Draw a picture.
Write the sum.

Check children's drawings.

1. 2 birds sit.
   3 birds eat.
   How many in all?

   $2 + 3 =$ ___5___ birds

2. 2 dogs run.
   2 dogs jump.
   How many in all?

   $2 + 2 =$ ___4___ dogs

3. 5 cows eat.
   1 cow sleeps.
   How many in all?

   $5 + 1 =$ ___6___ cows

Mark your answer.

4. 3 frogs hop.
   3 frogs sit.
   How many in all?

   4   5   6
   ○   ○   ⬤

5. 3 cats sit.
   2 cats sleep.
   How many in all?

   4   5   6
   ○   ⬤   ○

Reading Strategy • Use Pictures Clues

Using pictures can help you solve problems.

2 planes on the ground.
3 planes in the air.
How many in all?

Look at the
picture.

1. Count the planes **on the ground.**

____2.____ planes

2. Count the planes **in the air.**

____3____ planes

3. Write the addition sentence.

__2__ + __3__ = __5__ planes

Solve.

4. 2 jets on the ground.
   2 jets in the air.
   How many in all?

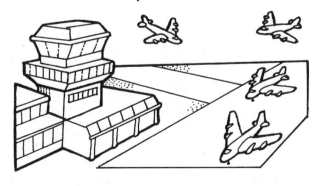

__2__ + __2__ = __4__ jets

5. 3 planes on the ground.
   1 plane in the air.
   How many in all?

__3__ + __1__ = __4__ planes

Modeling Subtraction Story Problems

Make up a subtraction story problem.
Write how many.

1.

__4__ balloons   __2__ pop   __2__ are left

2.

__5__ snow people   __1__ melts   __4__ are left

3.

__3__ toy boats   __3__ sink   __0__ are left

Make up a subtraction story problem.
How many are left? Mark your answer.

4.

○ 3   ● 4
○ 5   ○ 6

5.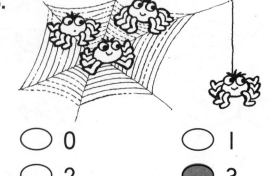

○ 0   ○ 1
○ 2   ● 3

Subtracting 1

Cross out 1.
Write how many are left. **Check children's work.**

1. There are 4 apples.
   Jon eats 1.

   How many are left?   4 – 1 = __3__ apples

2. There are 6 pears.
   Nancy eats 1.

   How many are left?   6 – 1 = __5__ pears

3. There are 2 apples.
   Nat eats 1.
   How many are left?

   ● 1   ◯ 2
   ◯ 3   ◯ 4

4. There are 3 oranges.
   Angel eats 1.
   How many are left?

   ◯ 0   ◯ 1
   ● 2   ◯ 3

PROBLEM SOLVING   PS19

Subtracting 2

Cross out pictures to show
the subtraction sentence.
Write how many are left. **Check children's work.**

1. 4 bugs are on a leaf.
   2 fly away.
   How many are left?

$4 - 2 =$ ___2___ bugs

---

2. 5 birds are on a fence.
   2 fly away.
   How many are left?

$5 - 2 =$ ___3___ birds

Choose the correct answer.

3. Which flag has 2 fewer
   than 6 stars?

4. Which animal has 2
   fewer than 4 legs?

Writing Subtraction Sentences

Draw a picture.
Write the number sentence.

Check children's drawings.

1. Pete has 5 balloons.
   3 blow away.
   How many are left?

   $5 - 3 = \underline{\ 2\ }$ balloons

2. Kathy has 6 balloons.
   She gives 3 away.
   How many are left?

   $6 - 3 = \underline{\ 3\ }$ balloons

3. Mia has 4 balloons.
   I pops.
   How many are left?

   $4 - 1 = \underline{\ 3\ }$ balloons

Which subtraction sentence tells how many are left?
Mark your answer.

4.

   ○ $5 - 4 = 1$
   ○ $5 - 3 = 2$
   ● $5 - 2 = 3$
   ○ $5 - 1 = 4$

5.

   ○ $4 - 0 = 4$
   ○ $4 - 1 = 3$
   ○ $4 - 3 = 1$
   ● $4 - 4 = 0$

Reading Strategy • Use Word Clues

Read the problem.
Look for word clues.
Solve the problem.

1. 4 dogs play.
   2 **run away**.
   How many are left?

There are __2__ dogs left.

2. 4 dogs play.
   2 **more** come.
   How many in all?

There are __6__ dogs in all.

Solve.

3. 6 kittens in a basket.
   3 get out.
   How many are left?

There are __3__ kittens left.

4. I puppy sleeps.
   5 more come.
   How many in all?

There are __6__ puppies in all.

Order Property

Write a number sentence to solve
each problem.

1. Lee has 2 green pens.
She buys 1 red pen.
How many pens does
Lee have?

$$\underline{2} + \underline{1} = \underline{3}$$

2. Tim has 1 green pen.
He buys 2 red pens.
How many pens does
Tim have?

$$\underline{1} + \underline{2} = \underline{3}$$

3. Sal draws 1 red star.
She draws 3 blue stars.
How many stars does
Sal draw?

$$\underline{1} + \underline{3} = \underline{4}$$

4. Laura draws 3 red stars.
She draws 1 blue star.
How many stars does
Laura draw?

$$\underline{3} + \underline{1} = \underline{4}$$

Choose the two number sentences
that show you can add in any order.

5. ⬤ $3 + 1 = 4$
$1 + 3 = 4$

○ $3 + 1 = 4$
$2 + 2 = 4$

○ not here

6. ○ $1 + 1 = 2$
$2 + 2 = 4$

○ $2 + 4 = 6$
$3 + 3 = 6$

⬤ $2 + 4 = 6$
$4 + 2 = 6$

Addition Combinations

Draw a picture.
Write the number sentence.

Check children's drawings.

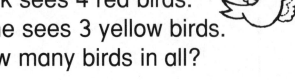

1. Jack sees 4 red birds.
   Jane sees 3 yellow birds.
   How many birds in all?

   __4__ + __3__ = __7__ birds

2. 3 birds eat seeds.
   5 more come.
   How many birds in all?

   __3__ + __5__ = __8__ birds

3. 1 bird takes a bath.
   6 more come.
   How many birds in all?

   __1__ + __6__ = __7__ birds

Mark the correct answer.

4. Which is a way to
   make 7?

   ○ 3 + 2
   ○ 2 + 4
   ● 0 + 7
   ○ 1 + 7

5. Which is a way to
   make 8?

   ● 4 + 4
   ○ 6 + 1
   ○ 2 + 4
   ○ 3 + 4

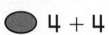

Harcourt Brace School Publishers

More Addition Combinations

Draw a picture.
Write the number sentence.

Check children's drawings.

1. Stan reads 4 books.
Tony reads 5 books.
How many books in all?

   __4__ + __5__ = __9__ books

2. Mia checks out 2 books.
Zack checks out 6 books.
How many books in all?

   __2__ + __6__ = __8__ books

3. 7 books are on the shelf.
2 books are on the desk.
How many books in all?

   __7__ + __2__ = __9__ books

Mark the correct answer.

4. The sum of two numbers is 9. Which are the two numbers?

   ○ 4 and 4
   ● 3 and 6
   ○ 7 and 1
   ○ 3 and 5

5. There are 10 children at a party. Which tells how many girls and boys there are?

   ○ 3 girls and 3 boys
   ○ 5 girls and 4 boys
   ○ 3 girls and 6 boys
   ● 7 girls and 3 boys

Horizontal and Vertical Addition

Write the problem two ways.

1. Ned has 6 fish.
He buys 4 more.
How many fish in all?

__10__ fish

$6 + 4 = 10$

$$\begin{array}{r} 6 \\ + 4 \\ \hline 10 \end{array}$$

---

2. Alice has 3 pennies.
She gets 6 more.
How many pennies in all?

__9__ pennies

$3 + 6 = 9$

$$\begin{array}{r} 3 \\ + 6 \\ \hline 9 \end{array}$$

Mark the correct answer.

3. Pam writes this addition sentence.

$$3 + 5 = 8$$

Which problem is the same?

○ $\begin{array}{r} 3 \\ +4 \\ \hline 7 \end{array}$ ● $\begin{array}{r} 3 \\ +5 \\ \hline 8 \end{array}$ ○ $\begin{array}{r} 4 \\ +4 \\ \hline 8 \end{array}$

4. Ali writes this problem.

$$\begin{array}{r} 4 \\ +5 \\ \hline 9 \end{array}$$

Which addition sentence is the same?

● $4 + 5 = 9$

○ $4 + 4 = 8$

○ $5 + 2 = 7$

○ $6 + 3 = 9$

Reading Strategy • Using Pictures

Using pictures can help you solve problems.

Tammy buys a top.
She buys a doll.
How much does she spend?

1. Look at the pictures.
   Write the answer.
   How much does the top cost? __4__ ¢

   How much does the doll cost? __6__ ¢

2. Write an addition sentence.
   Solve the problem.

   __4__ ¢ + __6__ ¢ = __10__ ¢

Solve.

3. Jim buys a boat.
   He buys a car.
   How much does
   he spend?

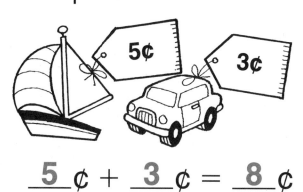

   __5__ ¢ + __3__ ¢ = __8__ ¢

4. Amy buys a ball.
   She buys jacks.
   How much does she
   spend?

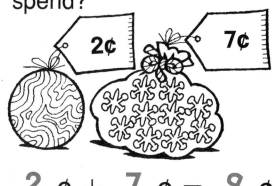

   __2__ ¢ + __7__ ¢ = __9__ ¢

Counting On 1 and 2

Count on to add.
Write the sum.

1.

6 birds in a house.
I more comes.
How many birds in all?

$6 + 1 = \underline{7}$ birds

2.

8 bees in a hive.
2 more come.
How many bees in all?

$8 + 2 = \underline{10}$ bees

3.

3 bears in a cave.
2 more come.
How many bears in all?

$3 + 2 = \underline{5}$ bears

Mark the correct answer.

4. Which has same sum?

$1 + 2 = \underline{}$

○ $1 + 1$
● $2 + 1$
○ $3 + 1$
○ $4 + 1$

5. Which has same sum?

$4 + 2 = \underline{}$

○ $4 + 1$
● $3 + 3$
○ $5 + 0$
○ $2 + 2$

Harcourt Brace School Publishers

Counting On 3

Count on to add.
Write the sum.

1.   Jan has 4 pennies.
She gets 3 more.
How many pennies in all?

$4 + 3 = \underline{\quad 7 \quad}$ pennies

2.   Dave has 8 pennies.
He gets 2 more.
How many pennies in all?

$8 + 2 = \underline{\quad 10 \quad}$ pennies

3.   Rick 6 pennies.
He gets 3 more.
How many pennies in all?

$6 + 3 = \underline{\quad 9 \quad}$ pennies

Mark the correct answer.

4. Which is a way to make 4?

- ⬤ $3 + 1$
- ◯ $3 + 2$
- ◯ $1 + 2$
- ◯ $4 + 2$

5. Which is a way to make 8?

- ◯ $5 + 1$
- ⬤ $5 + 3$
- ◯ $5 + 2$
- ◯ $5 + 4$

Doubles

Write a number sentence.

Check children's drawings.

1. Annie has 5 crayons.
   Max has the same number.
   How many do they have in all?

   __5__ + __5__ = __10__ crayons

2. Jill has 4 pens.
   Nick has 4 more than Jill.
   How many does Nick have?

   __4__ + __4__ = __8__ pens

Mark the correct answer.

3. Which doubles fact goes
   with the picture?

- ● 1 + 1 = 2
- ○ 2 + 2 = 4
- ○ 3 + 3 = 6
- ○ 4 + 4 = 8

4. Which doubles fact goes
   with the picture?

- ○ 2 + 2 = 4
- ○ 3 + 3 = 6
- ● 4 + 4 = 8
- ○ 5 + 5 = 10

Addition Facts Practice

Draw a picture.
Solve.

Harcourt Brace School Publishers

Check children's drawings.

1. A bike has 2 wheels.
A car has 4 wheels.
How many wheels in all?

____6____ wheels

2. A wagon has 4 wheels.
How many wheels do
2 wagons have?

____8____ wheels

3. A truck has 6 wheels.
A van has 4 wheels.
How many wheels in all?

____10____ wheels

Mark the correct answer.

4. Which is a doubles fact?

○ $2 + 3 = 5$

⬤ $2 + 2 = 4$

○ $4 + 2 = 6$

5. Which is **not** a doubles fact?

⬤ $4 + 3 = 7$

○ $3 + 3 = 6$

○ $4 + 4 = 8$

PROBLEM SOLVING   PS31

Reading Strategy • Use Word Clues

Using word clues can help
you solve problems.

Read the problem.
Look for word clues.
Draw a picture.
Solve.

Check children's drawings.

1. Tom has **6 plants.**
   He **gives 2 away.**
   How many are left?

   There are ___4___ plants left.

2. **3 plants** are **big.**
   2 plants are **little.**
   How many plants in all?

   ___5___ plants

3. Toby has 2 big plants.
   He buys 4 little plants.
   How many plants in all?

   ___6___ plants

4. Mike picks 5 flowers.
   He gives 3 away.
   How many are left?

   ___2___ flowers

Harcourt Brace School Publishers

Name _____

Subtraction Combinations

Draw a picture.
Subtract.

Check children's drawings.

1. There are 8 carrots.
   Beth eats 3.
   How many carrots are left?
    ___5___ carrots

2. There are 7 plums.
   Steve picks 4.
   How many plums are left?
   ___3___ plums

3. There are 8 beets.
   Lynn eats 5.
   How many beets are left?
   ___3___ beets

Mark the correct answer.

4. Which subtraction
   sentence tells how
   many are left?

   ◯ 6 − 3 = 3
   ◯ 3 − 3 = 0
   ⬤ 4 − 3 = 1
   ◯ 3 − 2 = 1

5. Which subtraction
   sentence tells how
   many are left?

   ◯ 2 − 1 = 1
   ⬤ 4 − 1 = 3
   ◯ 3 − 1 = 2
   ◯ 4 − 4 = 0

PROBLEM SOLVING   PS33

More Subtraction Combinations

Draw a picture.
Subtract.

1. Jean has 10 pennies.
   She spends 6.
   How many pennies are left?
    __4__ pennies

2. Jim has 7 pennies.
   He spends 2.
   How many pennies are left?
    __5__ pennies

3. Nelda has 9 pennies.
   She spends 3.
   How many pennies are left?
    __6__ pennies

Mark the correct answer.

4. Which comes next?

$$9 - 0 = \underline{\hspace{1cm}}$$
$$9 - 1 = \underline{\hspace{1cm}}$$
$$9 - 2 = \underline{\hspace{1cm}}$$

○ $9 - 6 = \underline{\hspace{1cm}}$
○ $9 - 5 = \underline{\hspace{1cm}}$
○ $9 - 4 = \underline{\hspace{1cm}}$
● $9 - 3 = \underline{\hspace{1cm}}$

5. Which comes next?

$$7 - 0 = \underline{\hspace{1cm}}$$
$$7 - 1 = \underline{\hspace{1cm}}$$
$$7 - 2 = \underline{\hspace{1cm}}$$

● $7 - 3 = \underline{\hspace{1cm}}$
○ $7 - 4 = \underline{\hspace{1cm}}$
○ $7 - 5 = \underline{\hspace{1cm}}$
○ $7 - 6 = \underline{\hspace{1cm}}$

Harcourt Brace School Publishers

Vertical Subtraction

Write the problem two ways.

1. Ken sees 5 fish.
   He sees 2 swim away.
   How many are left?

   ___3___ fish   5 − 2 = 3

   $$\begin{array}{r} 5 \\ -2 \\ \hline 3 \end{array}$$

2. Amy sees 10 butterflies.
   She sees 7 fly away.
   How many are left?

   ___3___ butterflies  10 − 7 = 3

   $$\begin{array}{r} 10 \\ -7 \\ \hline 3 \end{array}$$

Mark the correct answer.

3. Jody writes this
   subtraction sentence.

   $$7 - 3 = 4$$

   Which problem is the
   same?

   ● $\begin{array}{r} 7 \\ -3 \\ \hline 4 \end{array}$ ○ $\begin{array}{r} 7 \\ -4 \\ \hline 3 \end{array}$

4. Lou writes this problem.

   $$\begin{array}{r} 7 \\ -2 \\ \hline 5 \end{array}$$

   Which subtraction
   sentence is the same?

   ● $7 - 2 = 5$
   ○ $7 - 0 = 7$
   ○ $7 - 5 = 2$
   ○ $5 - 2 = 5$

Fact Families

Write number sentences.

1. Mr. Hill gives Ann these numbers. What fact family can she write?

4   l   5

$$\underline{4} + \underline{l} = \underline{5}$$
$$\underline{l} + \underline{4} = \underline{5}$$
$$\underline{5} - \underline{4} = \underline{l}$$
$$\underline{5} - \underline{l} = \underline{4}$$

2. Mr. Hill gives Chris these numbers. What fact family can he write?

3   2   l

$$\begin{array}{cccc} 2 & l & 3 & 3 \\ +\,l & +\,2 & -\,2 & -\,l \\ \hline 3 & 3 & l & 2 \end{array}$$

Mark the correct answer.

3. What are the numbers in this fact family?

$$4 + 4 = 8$$
$$8 - 4 = 4$$

- ● 4, 8
- ○ l, 4, 8
- ○ 0, 4, 8

4. What are the numbers in this fact family?

$$4 + 3 = 7$$
$$3 + 4 = 7$$
$$7 - 3 = 4$$
$$7 - 4 = 3$$

- ○ 4, 7, ll
- ○ 3, 4, 5
- ● 4, 3, 7

Harcourt Brace School Publishers

Subtracting to Compare

Draw a picture.
Subtract to compare.

1. Mia has 4 bowls.
   She has 2 fish.
   How many more bowls
   than fish does she have?

   ___2___ more bowls

2. Rich has 5 dogs.
   He has 4 bones.
   How many more bones
   does he need?

   ___1___ more bone

Mark the correct answer.

3. Which number sentence
   goes with this picture?

   ○ 4 + 3 = 7
   ○ 3 + 4 = 7
   ● 7 − 3 = 4
   ○ 7 − 7 = 0

4. Which number sentence
   goes with this picture?

   ○ 4 + 3 = 7
   ○ 3 + 4 = 7
   ● 4 − 3 = 1
   ○ 7 − 3 = 4

Counting Back 1 and 2

Use the number line.
Count back to subtract.

1. A robin is on number 8.
   It takes 1 hop back.
   What number is it on?

   __7__

---

2. A bluebird is on number 10.
   It takes 1 hop back.
   What number is it on?

   __9__

---

3. A crow is on number 7.
   It takes 2 hops back.
   What number is it on?

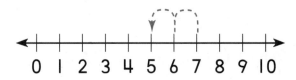

   __5__

Mark the correct answer.

4. Which number sentence
   does this number line
   show?

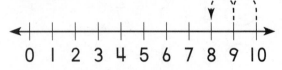

   ○ $10 - 1 = 9$
   ⬤ $10 - 2 = 8$

5. Which number sentence
   does this number line
   show?

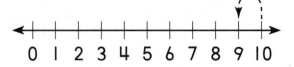

   ⬤ $10 - 1 = 9$
   ○ $10 - 3 = 7$

Harcourt Brace School Publishers

Counting Back 3

Use the number line.
Count back to subtract.

1. Fred stands on number 8.
   He takes 3 hops back.
   What number is he on now?

   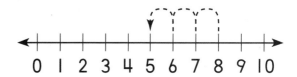

   <u>5</u>

2. Megan stands on number 4.
   She takes 2 hops back.
   What number is she on now?

   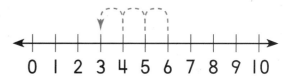

   <u>2</u>

3. Winnie stands on number 6.
   She takes 3 hops back.
   What number is she on now?

   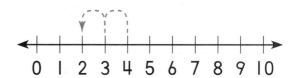

   <u>3</u>

Mark the correct answer.

4. Farley has 3 pennies.
   He finds 1 more.
   How many pennies does
   Farley have in all?

   ○ 1 penny
   ○ 2 pennies
   ● 4 pennies

5. Brooks has 3 pennies.
   He loses 1 penny.
   How many pennies does
   Brooks have left?

   ● 2 pennies
   ○ 3 pennies
   ○ 4 pennies

Subtracting Zero

Draw a picture.
Write the subtraction sentence.

Check children's drawings.

1. There are 4 crackers.
   Ned eats 4.
   How many are left?

   __4__ – __4__ = __0__ crackers

2. There are 6 oranges.
   Liza eats 0.
   How many are left?

   __6__ – __0__ = __6__ oranges

Mark the correct answer.

3. Which number sentence
   tells this story?

   ○ 6 – 1 = 5
   ⬤ 6 – 6 = 0
   ○ 6 – 0 = 6
   ○ not here

4. Which number sentence
   tells this story?

   ○ 6 – 1 = 5
   ○ 6 – 6 = 0
   ⬤ 6 – 0 = 6
   ○ not here

Harcourt Brace School Publishers

Facts Practice

Draw a picture.
Write the number sentence.

Check children's drawings.

1. Carla has 7 leaves.
   She drops 3.
   How many leaves are left?

__7__ − __3__ = __4__ leaves

2. Jim has 4 shells.
   He finds 2 more.
   How many shells in all?

__4__ + __2__ = __6__ shells

3. Pete has 8 nuts.
   He finds 1 more.
   How many nuts in all?

__8__ + __1__ = __9__ nuts

Mark the correct answer.

4. Which number names
   the sum?

   8 + 2 = _____

   ⬤ 10
   ◯ 9
   ◯ 8

5. Which number names the
   difference?

   10 − 2 = _____

   ◯ 10
   ◯ 9
   ⬤ 8

Reading Strategy • Use Word Clues and Pictures

Using word clues and pictures can help you
solve problems.

Read the problem.
Look for word clues.
Look at the picture.
Cross out the frogs that hop away.
Solve.

1. 5 frogs sit on a log.
   2 **hop away**.
   How many now?

   $5 - 2 = \underline{3}$ frogs

Solve.

2. 6 fish swim in a group.
   4 swim away.
   How many now?

   _____ 2 fish

3. 3 crabs sit on the bottom.
   3 more come.
   How many now?

   _____ 6 crabs

Harcourt Brace School Publishers

Solid Figures

Color the toys.

1. Tina wants to play
   with spheres.
   Color them red.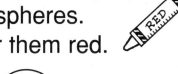

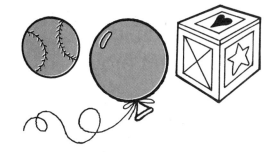

2. Matt wants to play
   with cones.
   Color them blue.

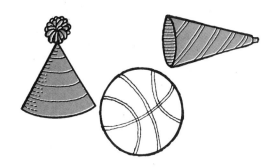

3. Luis wants to play
   with rectangular prisms.
   Color them yellow.

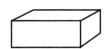

Which solid figure matches? Mark the answer.

4.

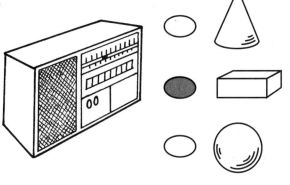

5.

More Solid Figures

Color the objects.

1. Look for things shaped
   like a pyramid.
   Color them brown.

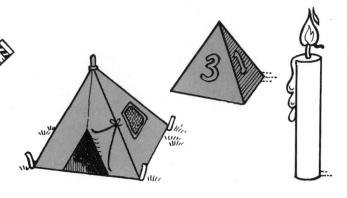

---

2. Look for things shaped
   like a cylinder.
   Color them green.

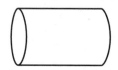

---

3. Look for things shaped
   like a cube.
   Color them red.

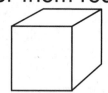

Which solid figure matches? Mark the answer.

4.

5.

Harcourt Brace School Publishers

Sorting Solid Figures

Circle the correct figures.

1. Ned wants to stack some blocks. Which blocks should he choose?

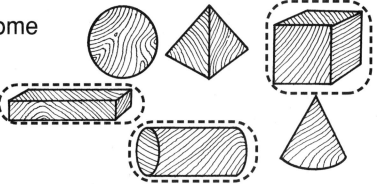

2. Jenny wants to slide blocks down a ramp. Which blocks should she choose?

3. Tom wants to roll some toys down a ramp. Which toys should he choose?

4. Which figure will roll and stack?

5. Which figure will slide but **not** roll?

More Sorting Solid Figures

Color the correct figures.

1. Pat builds a fence. She uses blocks with 2 faces. Which kind does she use?

2. Toby adds tops to some towers. He uses blocks with 1 face or 5 faces. Which kinds does he use?

3. Rita builds a house. She uses blocks that will slide and stack. Which kinds does she use?

Mark the correct answer.

4. How many faces?

 4

 5

 6

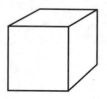

5. How many faces?

 0

 1

 2

Reading Strategy • Use Picture Clues

Using picture clues can
help you solve problems.

1. Look at the model.
   Count the cubes you see.

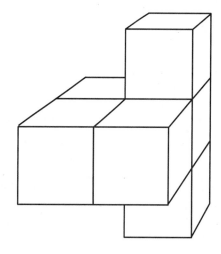

---

2. Look for hidden cubes.
   Count them.
   Write how many cubes.

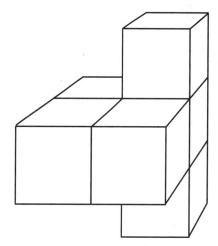

___6___ cubes

Write how many cubes.

3.

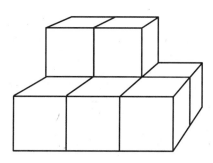

___8___ cubes

4.

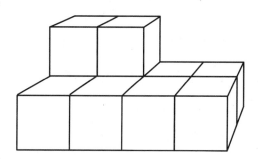

___10___ cubes

Plane Figures

Draw the correct shape.
Write its name.

 circle   **square**    **rectangle**    **triangle**

1. Nan draws around a face.
   What figure is the face?

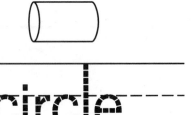

circle

2. Tim draws around a face.
   What figure is the face?

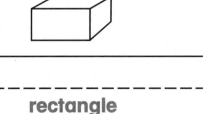

rectangle

3. Nate draws around a face.
   What figure is the face?

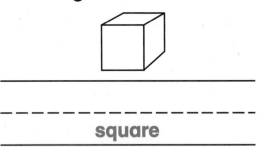

square

4. Pam draws around a face.
   What figure is the face?

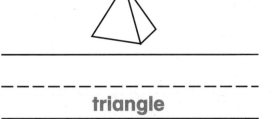

triangle

Mark the correct answer.

5. A cube has 6 faces
   shaped like me.
   What am I?

   ● square   ○ rectangle
   ○ triangle   ○ circle

6. A cylinder has 2 faces
   shaped like me.
   What am I?

   ○ square   ○ rectangle
   ○ triangle   ● circle

Harcourt Brace School Publishers

Sorting Plane Figures

Draw the correct shape.

 circle

 square

 **rectangle**

 **triangle**

1. Patty chooses a shape with 0 sides and 0 corners.	
2. Denny chooses a shape with 3 sides and 3 corners.	
3. Renee chooses a shape that is a face on a cube.	

Mark the correct answer.

4. Lee drew a closed figure with 6 sides.
How many corners does it have?

○ 4 corners

● 6 corners

○ 8 corners

5. Don drew a closed figure with 4 sides and 4 corners. Which figure did he draw?

● square

○ circle

○ triangle

Congruence

Draw a figure.   Check children's drawings. Answers may vary.

1. Karen makes a toy cat.
   She draws this pattern
   for the cat's ears.
   Draw a figure that is the
   same size and shape.

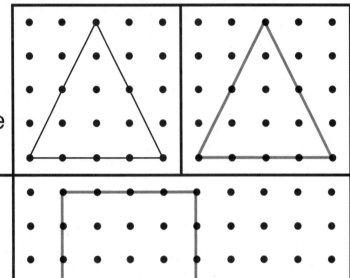

2. Ben makes a clown
   puppet.
   He traces around a
   cube for the buttons.
   Show what he draws.

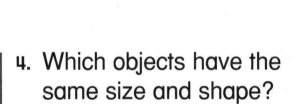

3. These figures are _____.

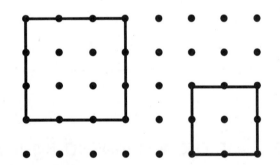

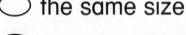
○ the same size
● the same shape
○ the same size and
   shape

4. Which objects have the
   same size and shape?

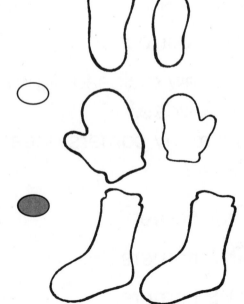

Symmetry

Solve.

1. Ruth wants leaves that have 2 sides that match. Draw lines to show the ones she should keep.

2. Dave looks for shells that have the same size and shape. Color the ones he should keep.

3. Polly looks for flowers that have 2 sides that match. Draw lines to show the ones she should keep.

4. Steve looks for stamps that have 3 corners and 3 sides. Color the ones he should keep.

Mark the correct answer.

5. How many ways can you divide this kite into two sides that match?

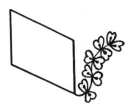

   ○ I way

   ● 2 ways

   ○ 3 ways

6. On which word can you draw a line to make two sides that match?

   ● WOW

   ○ WILL

   ○ WON

Open and Closed

Draw the figure.
Write **open** or **closed**.

Check children's drawings.

1. Tony draws a figure.
   It has 4 sides and 3 corners.

   _____

   This figure is ____**open**____.

2. Nick draws a figure.
   It has 7 sides and 7 corners.

   _____

   This figure is ____closed____.

3. Sally draws a figure.
   It has 6 sides and 5 corners.

   _____

   This figure is ____open____.

Mark the correct answer.

4. Which name begins with a closed letter?

   ◯ Carol
   ⬤ Bob
   ◯ Mary
   ◯ Saul

5. Which name begins with an open letter?

   ⬤ Jack
   ◯ Doug
   ◯ Oliver
   ◯ Bill

Inside, Outside, On

Draw a picture.
Solve.

Check children's drawings.

1. Tom draws a nest.
He draws 2 birds **inside** the nest.
He draws 1 bird **on** the nest.
He draws 2 birds **outside**
the nest. How many birds in all?

____5____ birds

2. Simon draws a rectangle.
He draws a circle **inside**
the rectangle.
He draws a square that is
outside the circle but
inside the rectangle.
Show what Simon draws.

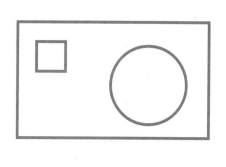

Mark the correct answer.

3. Where is the man?

○ inside the car

○ on the car

● outside the car

4. Where is
the bird?

● inside the birdhouse

○ on the birdhouse

○ outside the birdhouse

Reading Strategy • Use Word Clues

Use the position words **left** and
right to find where the friends will meet.

1. Find the X.
2. Walk right until you see the bench.
3. Turn and walk to the right of the slide.
4. Turn left at the tree.
5. Walk until you see me.

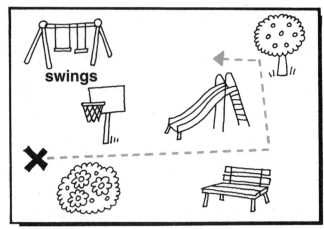

swings

left   **right**

1. Read all the directions.
   Look for the position words **left** and **right**.

2. Follow each direction. Use the position words.
   Mark the path you would take.

3. Solve the problem. Where will the friends meet?

   _____

   _

                    at the swings

Solve.

4. Draw a picture of your playground.
   Draw yourself to the **right** of your favorite place to play.

Check children's drawings.

Harcourt Brace School Publishers

Name _____

Positions on a Grid

Draw pictures on
the grid.

Start at the ☆.

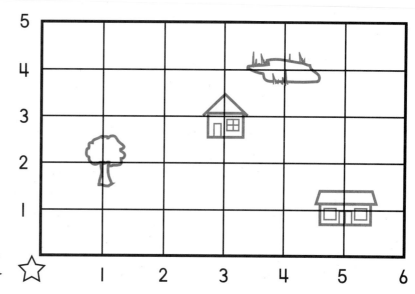

↑ **Up**

Right →

1. Lin lives in a house 3
   spaces to the right
   and 3 spaces up.

   Draw a 🏠 .

2. A tree grows I space to
   the right and 2 spaces

   up. Draw a 🌳 .

3. Lin likes to fish in a pond.
   It is 4 spaces to the right
   and 4 spaces up.

   Draw a 🐟 .

4. Lin helps her dad shop at
   a store. It is 5 spaces to
   the right and I space up.

   Draw a 🏪 .

Use the grid.
Mark the correct answer.

5. Which is to the left of
   Lin's house?

6. Which is farthest to
   the right?

Identifying Patterns

Draw to continue the pattern. **Check children's drawings.**

1. Beth made a belt with this pattern.
   Continue the pattern.

2. Ken made a bracelet with this pattern.
   Continue the pattern.

3. Saul made a belt with this pattern.
   Continue the pattern.

Mark the correct answer.

4. What is next in this pattern?

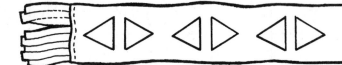

OO□OO□OO□OO□

   ⬭   ◯

   ◯   ▢

5. What is the rule for this pattern?

△O□△O□△O□

   ◯ triangle, circle

   ◯ square, circle, triangle

   ⬤ triangle, circle, square

Harcourt Brace School Publishers

Reproducing and Extending Patterns

Draw and color these patterns. **Check children's drawings.**

1. Marie uses red blocks to make the pattern triangle, square, circle. Draw her pattern two times.

2. Tom uses cubes to make the pattern yellow, red, red. Draw his pattern two times.

yellow   red   red   yellow   red   red

□ □ □ □ □ □

3. Jeff uses red squares and blue circles to make the pattern square, circle, circle. Draw his pattern two times.

Mark the correct answer.

4. What shape comes next?

-  ○ ○
-  □ ○
-  ○ □
-  □ □

5. Which shows the pattern square, triangle?

- ○ ○ □ △ △
-  □ △ △
- ○ □ ○ ○ △
- ○ △ ○ □ □

Making and Extending Patterns

Draw and color these patterns. **Check children's drawings.**

1. Glenn has 3 circles,
3 squares, and 3 triangles.
Show a pattern he can make.

---

2. Arlene has 3 circles and 6 cubes.
Show a pattern she can make.

---

3. Carol uses green circles, red squares,
and blue triangles to make the pattern
square, triangle, circle.
Draw her pattern three times.

Mark the correct answer.

4. Find a different pattern
that uses the same
shapes as this one.

○

●

○

5. Which shapes come next
in the pattern?

●

○

○

Harcourt Brace School Publishers

Reading Strategy • Make Predictions

Making predictions can help you solve problems.

Ruth makes this pattern with beads.
There is a mistake in the pattern.
What mistake do you see?

1. Look at the pattern.
   What is the pattern rule?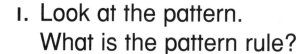

square, triangle, triangle

2. Use the pattern rule.
   Say the first 7 shapes
   in Ruth's pattern.

3. Make a prediction.
   What comes next in the pattern?
   Circle the mistake.
   Draw the correct pattern.

Find the pattern rule. Circle the mistake.
Make a prediction and continue the pattern.

4. Tim makes this block
   pattern. Circle the mistake.
   Draw the correct pattern.

5. Nora makes this bead
   pattern. Circle the mistake.
   Draw the correct pattern.

Counting On to 12

Draw how many come.
Write the sum.

Check children's drawings.

1. 8 children are in the
   playhouse. 2 more come.
   How many children in all?

$$\underline{8} + \underline{2} = \underline{10}$$

2. 8 children are in the sub.
   3 more come.
   How many children in all?

$$\underline{8} + \underline{3} = \underline{11}$$

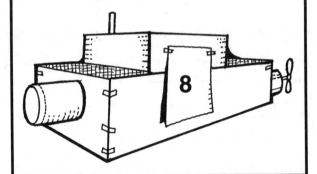

3. 9 children make a snow fort.
   3 more come.
   How many children in all?

$$\underline{9} + \underline{3} = \underline{12}$$

Mark the correct answer.

4. Count on.
   Which number is the sum?

   $$9 + 2 = \underline{}$$

   ● 11   ○ 13

   ○ 12   ○ 14

5. Count on. Which
   numbers have the
   sum 11?

   ○ 3 + 9   ● 9 + 2

   ○ 8 + 2   ○ 2 + 3

Harcourt Brace School Publishers

Doubles to 12

6 Pears   5 apples   4 plums   3 pineapples

Draw a picture.
Write the doubles facts.

Check children's drawings.

1. Mrs. Green buys 2 baskets of apples. How many apples does she buy?

   __5__ + __5__ = __10__ apples

2. Mr. Wills buys 2 baskets of pineapples. How many pineapples does he buy?

   __3__ + __3__ = __6__ pineapples

3. Ms. Young buys 2 baskets of plums. How many plums does she buy?

   __4__ + __4__ = __8__ plums

Mark the correct answer.

4. Which is a doubles fact?

   ○ 6 + 4 = 10
   ○ 6 + 5 = 11
   ● 6 + 6 = 12
   ○ not here

5. Which is not a doubles fact?

   ○ 3 + 3 = 6
   ○ 4 + 4 = 8
   ○ 5 + 5 = 10
   ● 6 + 5 = 11

Three Addends

Use ⬜. Draw them.
Write the sum.

Check children's drawings.

1. Ben scores 2 points.
   Marge scores 5 points.
   Lynn scores 5 points.
   How many points in all?

   __12__ points

2. Paul hits 4 runs.
   Mary hits the same number.
   How many runs in all?

   __8__ runs

Mark the correct answer.

3. Which team scored
   more points?

Lions	Tigers
2	1
3	4
4	3

○ Tigers

● Lions

4. Which sum is the same
   as 3 + 2 + 5?

○ 5 + 2 + 2

○ 3 + 3 + 2

● 8 + 0 + 2

○ 4 + 1 + 2

Practice the Facts

Draw a picture.
Write the number sentence.

Check children's drawings.

1. Carl spends 7¢.
   Jill spends 1¢.
   Bob spends 4¢.
   How much do they spend?

   __7__ + __1__ + __4__ = __12__ ¢

2. A pencil costs 8¢.
   A pen costs 2¢ more.
   How much does a pen cost?

   __8__ + __2__ = __10__ ¢

3. Pete spends 4¢.
   Bill spends the same amount.
   How much do they spend?

   __4__ + __4__ = __8__ ¢

Mark the correct answer.

4. Which prices have a sum
   of 10¢?

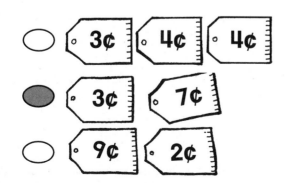

   ○ 3¢ 4¢ 4¢
   ● 3¢ 7¢
   ○ 9¢ 2¢

5. Lois has 10¢. Which toy
   could she buy?

   ○ 11¢
   ○ 12¢
   ● 9¢

Reading Strategy • Use Word Clues

Sometimes problems use repeated words.
Using **word clues** can help you read and solve problems.

8 fish hide.
2 more fish hide.
How many fish hide?

1. Look for word clues in the problem.
   What are the clue words?

   _____
   – – – – – – – – – – – –
   8 fish, 2 more fish

2. Draw a picture.
   Write an addition sentence to solve the problem.

   __8__ + __2__ = __10__

   __10__ fish

   Check children's drawings.

Solve.

3. 7 frogs sit on a rock.
   4 more frogs sit.
   How many frogs sit?

   __7__ + __4__ = __11__

   __11__ frogs

Harcourt Brace School Publishers

Relating Addition and Subtraction

Write a number sentence.
Solve.

1. Nora has 5 apples.
   She picks 2 more.
   How many apples does
   she have in all?

   $\underline{5}\;\oplus\;\underline{2}\;=\;\underline{7}$
   apples

2. Nora has 7 apples.
   She eats 2.
   How many apples
   does she have left?

   $\underline{7}\;\ominus\;\underline{2}\;=\;\underline{5}$
   apples

3. Bob has 9 plums.
   He eats 1.
   How many plums does
   he have left?

   $\underline{9}\;\ominus\;\underline{1}\;=\;\underline{8}$
   plums

Mark the correct answer.

4. The sum of two numbers
   is 9. One number is 6.
   What is the other number?

   ○ 2
   ● 3
   ○ 4
   ○ not here

5. The sum of two numbers
   is 12. One number is 5.
   What is the other number?

   ● 7
   ○ 6
   ○ 5
   ○ not here

Counting Back

Count back or count on to solve.

Check children's work.

1. Betty stands on number 6.
   She takes 2 hops back.
   What number is she on now?

   __4__

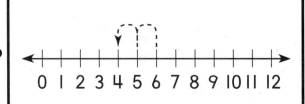

0 1 2 3 4 5 6 7 8 9 10 11 12

2. Ann stands on number 8.
   She takes 3 hops forward.
   What number is she on now?

   __11__

0 1 2 3 4 5 6 7 8 9 10 11 12

3. Stan stands on number 11.
   He takes 3 hops back.
   What number is he on now?

   __8__

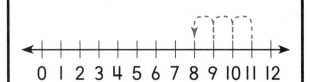

0 1 2 3 4 5 6 7 8 9 10 11 12

Mark the correct answer.

4. Count back.
   Which is the difference?

   $9 - 2 =$ _____

   ● 7
   ○ 8
   ○ 9
   ○ 10

5. Count on.
   Which is the sum?

   $9 + 2 =$ _____

   ○ 10
   ● 11
   ○ 12
   ○ 13

Compare to Subtract

Draw a picture.   Check children's drawings.
Solve.

1. There are 12 children in line.
   The bus has 8 seats.
   How many fewer seats than
   children are there?

   ___4___ fewer seats

2. There are 9 people waiting
   for the bus. The bus has
   7 empty seats.
   How many more people
   than seats are there?

   ___2___ more people

Mark the correct answer.

3. Which question goes with
   the problem?
   There are 8 children.
   4 children leave.

   ○ How many children
     in all?

   ● How many children
     are left?

   ○ How many more
     children are there?

4. Which question goes with
   the problem?
   There are 9 children.
   3 more children come.

   ● How many children
     in all?

   ○ How many children
     are left?

   ○ How many fewer
     children are there?

Fact Families

Draw a picture.
Write the number sentence.

Check children's drawings.

1. Martha has 5 red blocks.
She has 3 blue blocks.
How many blocks does
Martha have?

$$\underline{5} \ \textcircled{+} \ \underline{3} = \underline{8}$$
blocks

2. Toby has 8 cars.
He gives 3 cars to his sister.
How many cars does Toby
have left?

$$\underline{8} \ \textcircled{-} \ \underline{3} = \underline{5}$$
cars

3. Jan lines up 6 number cards.
She counts back 3 cards
from 6. On what number
does she stop?

$$\underline{3}$$

Mark the correct answer.

4. Which number sentence
belongs in this fact family?

$$8 + 3 = 11$$

- ⚪ $8 + 4 = 12$
- ⚪ $8 + 2 = 10$
- ⚪ $11 - 4 = 7$
- ⬤ $11 - 3 = 8$

5. Which number sentence
belongs in this fact family?

$$5 + 5 = 10$$

- ⚪ $10 - 0 = 10$
- ⬤ $10 - 5 = 5$
- ⚪ $6 + 6 = 12$
- ⚪ $6 + 4 = 10$

Harcourt Brace School Publishers

Reading Strategy • Reread

Rob has 8 forks.
He has 12 plates.
How many fewer forks than
plates does Rob have?

1. Read the problem.
   Write a number sentence.

$$12 - 8 = \underline{}$$

2. Reread the problem.
   Make sure the numbers you wrote
   match the numbers in the problem.

   8 forks

   12 plates

3. Look for words that tell you what to do.
   The words **how many fewer** tell you to
   compare two groups and subtract.

4. Solve the problem.

   $12 - 8 = \underline{4}$ fewer forks

Solve.

5. Rob has 11 straws.
   He has 9 cups.
   How many more straws
   than cups are there?

   $\underline{2}$ more straws

6. Rob has 12 bows.
   He has 7 balloons.
   How many fewer balloons
   than bows are there?

   $\underline{5}$ fewer balloons

Tens

Draw a picture.
Write how many groups of ten.

Check children's drawings.

1. Paul has 40 apples. How many
   bags of 10 apples can he make?

   ___4___ bags of 10

2. Nick has 60 pears. How many
   bags of 10 pears can he make?

   ___6___ bags of 10

3. Rosa has 80 nuts. How many
   bags of 10 nuts can she make?

   ___8___ bags of 10

How many counters?
Mark the correct answer.

4.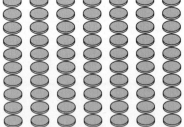

   ○ 5 tens = 50
   ○ 6 tens = 60
   ● 7 tens = 70
   ○ 8 tens = 80

5.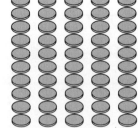

   ○ 4 tens = 40
   ● 5 tens = 50
   ○ 6 tens = 60
   ○ 7 tens = 70

Harcourt Brace School Publishers

Tens and Ones to 20

Draw a picture.
Write the number.

Check children's drawings.

1. Jack has 10 stickers. He finds 4 more. How many stickers does he have?

   ___14___ stickers

2. Fran has 60 stickers. She puts 10 stickers on each page. How many pages does she fill?

   ___6___ pages

3. Tonya has 10 stickers. She finds 8 more. How many stickers does she have?

   ___18___ stickers

How many tens and ones?
Mark the correct answer.

4.

   ○ 0 tens 3 ones
   ○ I ten 0 ones
   ● I ten 3 ones
   ○ 0 tens I one

5.

   ● I ten 7 ones
   ○ I ten 6 ones
   ○ I ten 8 ones
   ○ 0 tens 7 ones

Tens and Ones to 50

Draw a picture.
Solve.

1. Mrs. Hall has 3 boxes of 10 cookies. She buys 4 more cookies. How many cookies in all?

   __**34**__ cookies

2. Polly has 4 rolls of 10 mints. She buys 6 more mints. How many mints in all?

   __**46**__ mints

3. Suzy has 1 package of 10 crackers. She buys 7 more crackers. How many crackers in all?

   __**17**__ crackers

How many? Mark the correct answer.

4.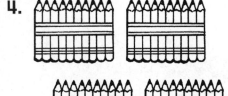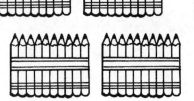

   ○ 33   ○ 34

   ● 43   ○ 44

5.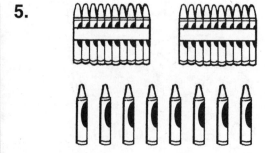

   ○ 18   ○ 23

   ○ 25   ● 28

Tens and Ones to 80

Draw a picture.
Write the number.

Check children's work.

1. Bob spins 6 tens and
   4 ones. What number
   does Bob spin?

   64

2. Marta spins 4 tens and
   9 ones. What number
   does Marta spin?

   49

3. Jimmy spins 7 tens and
   6 ones. What number
   does Jimmy spin?

   76

What is the number?
Mark the correct answer.

4. I have 6 tens.
   I have 8 ones.
   What number am I?

   ○ 61   ● 68

   ○ 81   ○ 86

5. I have 7 tens.
   I have 2 ones.
   What number am I?

   ○ 27   ○ 52

   ○ 70   ● 72

Harcourt Brace School Publishers

Tens and Ones to 100

Draw a picture.
Write the number.

1. Bill packed 8 boxes of
   10 books.
   He found 3 more books.
   How many books in all?

   __83__ books

2. Molly packed 4 boxes of
   10 cards.
   She found 9 more cards.
   How many cards in all?

   __49__ cards

Mark the correct answer.

3. What number does
   Chris show?

   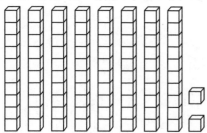

   ○ 28   ○ 72
   ● 82   ○ 83

4. What number does
   Laura show?

   ○ 77   ○ 79
   ○ 87   ● 97

Harcourt Brace School Publishers

PS74   PROBLEM SOLVING

Reading Strategy • Use Word Clues

Using word clues such as **more** and **fewer** can
help you use pictures to estimate.

Look at the sheep.
Which is the better estimate?

more than 10   fewer than 10

---

1. Read the problem. Think about the
   meaning of **more** and **fewer**.

more than 10   10   fewer than 10

---

2. Estimate the number of sheep.

Think: I group of 10 sheep and 3 more sheep.
There are **more than 10** sheep.

Circle the better estimate.

3.

more than 10

fewer than 10

4.

more than 10

fewer than 10

Ordinals

| first | second | third | fourth | fifth | sixth | seventh | eighth |

Use the picture to find the answer.

1. Lynn takes the seventh animal to school. What animal does she take?

_____
cat

2. Lynn counts all the animals that come after the third animal. How many are there?

5

3. Lynn adds a toy frog at the end of the line.
In what position is the frog?

_____
ninth

4. Lynn puts a toy cow between the duck and the hen. In what position is the hen now?

_____
sixth

Mark the correct answer.

5. The winner in a race crosses the finish line ____.

- ● first
- ○ second
- ○ third
- ○ fourth

6. There are nine seats left in a movie theater. You are the tenth in line. Will you get a seat?

- ○ yes
- ● no

Harcourt Brace School Publishers

Greater Than

Draw a picture.
Write the correct answer.

Check children's drawings.

1. Ann bakes 24 cookies.
   Leon bakes 36 cookies.
   Who bakes the greater number?

   Leon

2. Nick sells 12 cupcakes.
   Todd sells 21 cupcakes.
   Who sells the greater number?

   Todd

3. There are 3 people in line. Ted
   is behind Sally. Lou is behind
   Ted. Who is first in line?

   Sally

Mark the correct answer.

4. Which number is greater
   than 50?

   ○ 5   ⬤ 51
   ○ 15   ○ 50

5. Which number is greater
   than your age?

   ○ 0   ○ 4
   ○ 2   ⬤ 10

Less Than

Draw a picture.
Write the correct answer.

Check children's drawings.

1. Jean has 32 toy cars. Luke has 23 cars. Which number is less?

   _23_

2. Paul uses 65 blocks. Renee uses 56. Who uses the greater number of blocks?

   _____

   - - - - - - - - - - - - - - - - - -
              Paul
   _____

3. Marcy has 90 blocks. She makes 1 fence with 10 blocks. How many fences can she make?

   __9__ fences

Mark the correct answer.

4. Which number is less than 46?

    45
   ○ 46
   ○ 48
   ○ 49

5. Which number is less than 11?

    9
   ○ 19
   ○ 29
   ○ 39

Harcourt Brace School Publishers

Before, After, Between

Write the number.

1. Kay chooses a number between 75 and 77. What number does she choose?

   __76__

2. Which number is greater than 35?

   31   16   45   27

   __45__

3. Kerry finished the race before Tom. Sally finished the race after Tom. Who finished the race first?

   _____

   _____
   Kerry

Mark the correct answer.

4. Which number is just after 54?

   ● 55   ○ 52

   ○ 53   ○ 51

5. Which number is just before 40?

   ○ 29   ● 39

   ○ 35   ○ 41

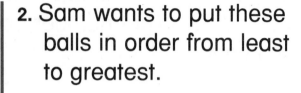

Order to 100

Look at the picture.
Write the numbers from least to greatest.

1. Mary wants to put these hats in order from least to greatest.

__12__ __22__ __62__ __82__

2. Sam wants to put these balls in order from least to greatest.

__60__ __63__ __67__ __70__

Write the correct answer.

3. Chet picks three number cards. The numbers are between 40 and 44. What numbers does he pick?

__41__ __42__ __43__

4. Nan picks three number cards. The numbers are after 34 but before 38. What numbers does she pick?

__35__ __36__ __37__

Mark the correct answer.

5. Which numbers are in order from least to greatest?

 12, 37, 39, 42

◯ 12, 39, 37, 42

◯ 42, 39, 37, 12

6. Which set of cards is in order from least to greatest?

◯ | 29 | 20 | 23 |

◯ | 23 | 20 | 29 |

 | 20 | 23 | 29 |

Harcourt Brace School Publishers

Counting by Tens

Draw a picture.
Count by tens.

1. There are 4 children in a family. Each child gets 10 dollars for a present. How much money do they get in all?

    _40_ dollars

2. Carlos has 7 packs of baseball cards. Each pack has 10 cards. How many cards does he have in all?

    _70_ cards

Mark the correct answer.

3. What three numbers come next in this pattern?

    20, 30, 40, ___, ___, ___

    ○ 41, 42, 43
    ○ 45, 50, 55
    ○ 60, 70, 80
    ● 50, 60, 70

4. How many in all?

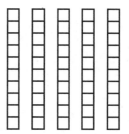

    ○ 5
    ○ 40
    ● 50
    ○ 60

Name _____

Counting by Fives

Draw a picture.
Skip-count to solve.

Check children's drawings.

1. Hannah is 6 years old today.
   Her grandmother gives her
   5 dollars for each year. How
   much money does Hannah get?

   __30__ dollars

2. Tory builds 8 towers.
   He uses 10 blocks in each tower.
   How many blocks does Tory use
   in all?

   __80__ blocks

Mark the correct answer.

3. How many in all?

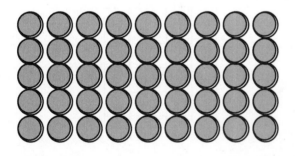

   ○ 9
   ○ 40
   ● 45
   ○ 90

4. What three numbers
   come next in this pattern?

   55, 60, 65, ___, ___, ___

   ○ 80, 85, 90
   ○ 75, 80, 85
   ○ 70, 80, 90
   ● 70, 75, 80

Counting by Twos

Draw a picture.
Skip-count to solve.

1. Doug thinks of a number.
It is between 35 and 40.
It is 2 more than 36.
What is Doug's number?

<u>38</u>

2. Sal thinks of a number.
It is between 70 and 80.
It is 5 more than 70.
What is Sal's number?

<u>75</u>

Mark the correct answer.

3. When you say every
second number, you are
counting by _____.

● twos
○ fives
○ tens

4. What three numbers
come next in this
pattern?

52, 54, 56, ___, ___, ___

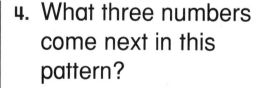

○ 57, 59, 61
● 58, 60, 62
○ 60, 65, 70

Even and Odd Numbers

Draw a picture to solve.
Then circle **odd** or **even**.

Check children's drawings.

1. Marissa plants 5 flowers.
   She plants 3 more. How many
   flowers does she plant?

   __8__ flowers   odd (even)

2. Bill mows 3 lawns on Monday.
   He mows double that number
   on Tuesday. How many lawns
   does he mow on both days?

   __9__ lawns   (odd) even

3. 6 children each plant 2 bulbs.
   How many bulbs do they
   plant in all?

   __12__ bulbs   odd (even)

Mark the correct answer.

4. The sum of an even
   number and an odd
   number is always _____.

   ○ even   ● odd

5. The sum of two odd
   numbers is always
   _____.

   ● even   ○ odd

PS84   PROBLEM SOLVING

Pennies and Nickels

Draw a picture. Count.
Write the amount.

Check children's drawings.

1. Lisa has 2 pennies.
   She finds 3 more pennies in a jar.
   How much money does Lisa
   have in all?

   __5__ ¢

2. Bob has 2 nickels.
   His mom gives him 1 more nickel.
   How much money does he have
   in all?

   __15__ ¢

Mark the correct answer.

3. Which group of coins is
   worth more?

   ◯ 6 pennies

   ◯ 2 nickels

   ⬤ 11 pennies

4. Hanna wants to count
   her coins. What should
   she say?

   ◯ 1¢, 2¢, 3¢, 4¢

   ◯ 5¢, 10¢, 15¢

   ⬤ 5¢, 10¢, 15¢, 20¢

Pennies and Dimes

Draw a picture. Count.
Write the amount.

Check children's drawings.

1. Ned has 4 dimes.
   He earns 2 more dimes.
   How much money does Ned
   have in all?

   <u>60</u> ¢

2. Patty has 2 nickels.
   She earns 3 more nickels.
   How much money does she
   have in all?

   <u>25</u> ¢

Mark the correct answer.

3. Mary has these coins.

How much money does
she have?

○ 2¢   ○ 10¢

○ 15¢   ● 20¢

4. Ben wants to count
   his coins. What should
   he say?

○ 1¢, 2¢, 3¢

○ 5¢, 10¢, 15¢

● 10¢, 20¢, 30¢

○ not here

Harcourt Brace School Publishers

Counting Collections of Nickels and Pennies

Draw a picture. Count.
Write the amount.

Check children's drawings.

1. Dan has 3 nickels in his bank.
   He puts in 4 pennies.
   How much money does Dan
   have in all?

   ___19___ ¢

2. Jessica has 2 dimes.
   Her aunt gives her 3 more dimes.
   How much money does
   Jessica have in all?

   ___50___ ¢

Mark the correct answer.

3. Which group of coins is
   worth the most?

   ○ 3 nickels and
      3 pennies

   ○ 2 nickels and
      9 pennies

   ○ 4 nickels and
      1 penny

   ● 3 nickels and
      7 pennies

4. Which amount do these
   coins add up to?

   ○ 10¢   ● 12¢

   ○ 22¢   ○ not here

Counting Collections of Dimes and Pennies

Draw a picture. Count.
Write the amount.

Check children's drawings.

1. Katie has 3 dimes.
   Her mom puts 2 more dimes
   in her lunch box. How much
   money does Katie have in all?

   __50__ ¢

2. John has 4 nickels in his bank.
   He puts in 4 pennies.
   How much money does he have
   in all?

   __24__ ¢

Mark the correct answer.

3. Which group of coins is
   worth the least?

   ◯ 12 pennies

   ● 1 dime 1 penny

   ◯ 1 dime 4 pennies

   ◯ 2 dimes

4. What amount do these
   coins add up to?

   ◯ 5¢   ◯ 13¢

   ● 23¢   ◯ 50¢

Harcourt Brace School Publishers

Reading Strategy • Use Pictures

Using pictures can help you
solve problems.

A toy truck costs 21¢.
Which coins could Thomas
use to buy the truck?

1. Look at the picture.
   Put the coins into groups.
   Then find the value for each group.

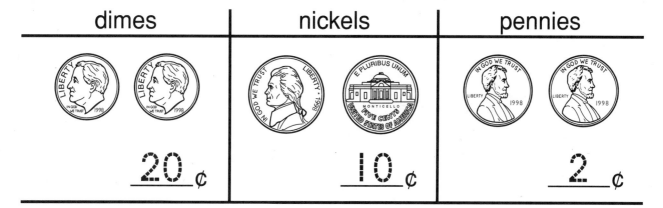

dimes	nickels	pennies
__20__ ¢	__10__ ¢	__2__ ¢

2. Write the number of coins from the groups
   that add up to the price of the toy truck.

   __1__ dime   __2__ nickels   __1__ penny = __21__ ¢

Solve.

3.

   A toy bus costs 12¢.
   Which coins could you
   use to buy it?

   __1__ dime
   __0__ nickels
   __2__ pennies

Trading Pennies, Nickels, and Dimes

Use the fewest coins.
Draw the coins.
Solve.

Check children's drawings.
Answers will vary.

1. Mrs. Polt needs dimes for parking. She has 20 pennies and 2 nickels. How many dimes can she trade for?

_____3_____ dimes

2. Mr. Miller uses dimes and nickels for tolls. He has 35 pennies. How many dimes and nickels can he trade for?

_____3_____ dimes   _____1_____ nickel

Mark the correct answer.

3. What are the fewest coins you can use to show 25¢?

○ 2 coins

● 3 coins

○ 5 coins

○ 25 coins

4. Tina has 4 coins.
José has 8 coins.
Who has more money?

○ Tina

○ José

● You cannot tell.

Harcourt Brace School Publishers

Name _____

Equal Amounts

Draw a picture.
Write the amount.

**Check children's drawings.
Answers will vary.**

1. Jill buys jacks that cost 30¢.
Show two ways she could pay.

__ dimes __ nickels __ pennies

__ dimes __ nickels __ pennies

2. Alex buys a ball that costs 45¢.
Show two ways he could pay.

__ dimes __ nickels __ pennies

__ dimes __ nickels __ pennies

Mark the correct answer.

3. Betty and Frank have the
same amount of money.
Betty has only dimes.
Frank has only nickels.
Who has fewer coins?

⬤ Betty

◯ Frank

◯ You cannot tell.

4. Joe and Carmen have
the same number of
coins. Do they have the
same amount of money?

◯ yes

◯ no

⬤ You cannot tell.

How Much Is Needed?

Draw a picture.
Solve.

Check children's drawings.

1. Nicky has 22¢ in her purse.
   She has 4 coins.
   What are they?

   2 dimes **0** nickels **2** pennies

2. Don has 1 dime and 7 nickels.
   Lisa has the same amount but
   fewer coins. What coins does
   she have?

   4 dimes **1** nickel

Mark the correct answer.

3. Ted uses fewer coins
   than Janna to show the
   same amount.
   This means that some of
   Ted's coins are worth

   _____.

    more

   ◯ less

   ◯ the same

4. A cookie costs 15¢. Joe
   uses the fewest coins to
   buy it. What coins does
   he use?

   ◯ 15 pennies

   ◯ 2 nickels and
   5 pennies

   ◯ 3 nickels

   ⬤ 1 dime and 1 nickel

Quarter

Draw a picture. Count.
Write the amount.

Check children's drawings.

1. Fran has 1 quarter.
She earns 1 dime and 1 nickel.
How much money does Fran
have in all?

___40___ ¢

2. Pete has 2 dimes.
He finds 1 more dime and 2
nickels in his bag. How much
money does Pete have in all?

___40___ ¢

Mark the correct answer.

3. Which group of coins is
worth the most?

- ◯ 1 quarter and
2 dimes
- ◯ 1 quarter and
3 nickels
- ⬤ 1 quarter and
5 nickels
- ◯ 1 quarter and
20 pennies

4. What amount do these
coins add up to?

- ◯ 36¢   ◯ 39¢
- ⬤ 41¢   ◯ 46¢

Reading Strategy • Make Predictions

Making predictions can help you
solve problems.

A toy plane costs 37¢. Charlie has
I quarter, I dime, I nickel, and
I penny. Does Charlie have
enough to buy the plane?

1. Use coins to show the
   amount Charlie has.
   Count two coins.

   <u>25</u> ¢, <u>35</u> ¢

2. Make a prediction.
   Do you think Charlie has
   enough to buy the plane? _____ **yes**

3. Continue counting Charlie's coins to find the value.

   <u>25</u> ¢, <u>35</u> ¢, <u>40</u> ¢, <u>41</u> ¢

Make a prediction. Then solve.
Use the fewest coins.

4. A toy bear costs 37¢.
   Jill has 2 dimes, 5 nickels,
   and 3 pennies. Can Jill
   buy the plane?
   If so, what coins should
   she use?

   <u>2</u> dimes

   <u>3</u> nickels

   <u>2</u> pennies

Ordering Months and Days

January	February	March	April	May	June
S M T W T F S	S M T W T F S	S M T W T F S	S M T W T F S	S M T W T F S	S M T W T F S
1 2 3	1 2 3 4 5 6 7	1 2 3 4 5 6 7	1 2 3 4	1 2 3 4 5 6 7 8 9	1 2 3 4 5 6
4 5 6 7 8 9 10	8 9 10 11 12 13 14	8 9 10 11 12 13 14	5 6 7 8 9 10 11	10 11 12 13 14 15 16	7 8 9 10 11 12 13
11 12 13 14 15 16 17	15 16 17 18 19 20 21	15 16 17 18 19 20 21	12 13 14 15 16 17 18	17 18 19 20 21 22 23	14 15 16 17 18 19 20
18 19 20 21 22 23 24	22 23 24 25 26 27 28	22 23 24 25 26 27 28	19 20 21 22 23 24 25	24/31 25 26 27 28 29 30	21 22 23 24 25 26 27
25 26 27 28 29 30 31		29 30 31	26 27 28 29 30		28 29 30

July	August	September	October	November	December
S M T W T F S	S M T W T F S	S M T W T F S	S M T W T F S	S M T W T F S	S M T W T F S
1 2 3 4	1	1 2 3 4 5	1 2 3	1 2 3 4 5 6 7	1 2 3 4 5
5 6 7 8 9 10 11	2 3 4 5 6 7 8	6 7 8 9 10 11 12	4 5 6 7 8 9 10	8 9 10 11 12 13 14	6 7 8 9 10 11 12
12 13 14 15 16 17 18	9 10 11 12 13 14 15	13 14 15 16 17 18 19	11 12 13 14 15 16 17	15 16 17 18 19 20 21	13 14 15 16 17 18 19
19 20 21 22 23 24 25	16 17 18 19 20 21 22	20 21 22 23 24 25 26	18 19 20 21 22 23 24	22 23 24 25 26 27 28	20 21 22 23 24 25 26
26 27 28 29 30 31	23/30 24/31 25 26 27 28 29	27 28 29 30	25 26 27 28 29 30 31	29 30	27 28 29 30 31

Use the calendar to answer each question.

1. Ann's birthday is in the third month of the year. In what month is her birthday?

2. Tom's birthday comes in the month after June. In what month is Tom's birthday?

_____

July

3. What is the month before December?

_____

November

4. What is the ninth month of the year?

_____

September

Mark the correct answer.

5. How many months are there in one year?

◯ 10 months

◯ 11 months

⬤ 12 months

◯ not here

6. What month comes between May and July?

◯ April

⬤ June

◯ August

◯ not here

Reading Strategy • Matching Text

Looking for matching words can help you solve
a problem.

Jon has a game on the second
Monday of November.
On what date is the game?

November						
SUN.	MON.	TUES.	WED.	THURS.	FRI.	SAT.
	1	2	3	4	5	6
7	8	9	10	11	12	13
14	15	16	17	18	19	20
21	22	23	24	25	26	27
28	29	30				

1. Read the problem.
   What date do you
   need to find? **the second Monday**

2. Match words from the problem with words
   on the calendar.
   Find **Monday** in the list of days at the top of
   the calendar.

3. Find the second Monday on the calendar.
   What is the date?

November 8

Use the calendar to answer the questions.

4. Linda's party is the third
   Tuesday of November.
   On what date is the party?

   _____

   November 16

5. Thanksgiving is the fourth
   Thursday of November.
   On what date is
   Thanksgiving?

   _____

   November 25

Ordering Events

Read each story.
Write **morning, afternoon,** or **evening** to tell
the time.

1. Luis is hungry after school. He eats an apple for a snack. Then he goes outside to play.

   afternoon

2. Sally takes too much time getting dressed. After breakfast, she must run to catch the school bus.

   morning

3. Marta takes the bus to her music teacher's house. She gets home before it is dark.

   afternoon

4. Tammy puts on her pajamas. Then she reads a story before going to sleep.

   evening

Mark the correct answer.

5. Which do most people do in the evening?

   ○ eat breakfast

   ○ eat lunch

   ● eat dinner

6. When does this happen?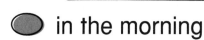

   ● in the morning

   ○ in the afternoon

   ○ in the evening

Reading Strategy • Use Prior Knowledge

How much time do these things take to do?
Number them from the shortest time to the longest time.

Leslie makes
her bed.

2

Leslie hangs
up her coat.

1

Leslie eats
dinner.

3

---

1. Read each sentence.
   About how long does it take you to do each thing?

   It takes a very short time to hang up a coat.
   It takes a longer time to make a bed.
   It takes the longest time to eat dinner.

Number these things from the shortest time
to the longest time. Use 1, 2, 3.

2. Mark gets ready for bed.
   He takes a bath.
   He brushes his hair.
   He brushes his teeth.

   3 He takes a bath.

   1 He brushes his
   hair.

   2 He brushes his
   teeth.

Reading the Clock

Use the clock.
Write the time two ways.

1. Alice catches the school
   bus at

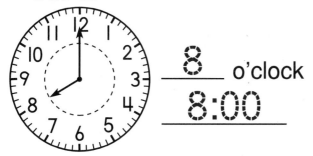

__8__ o'clock

8:00

2. John has math at

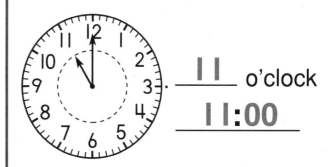

__11__ o'clock

11:00

3. Sam has gym at

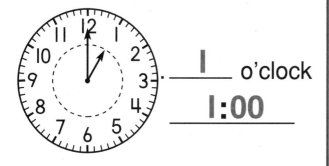

__1__ o'clock

1:00

4. Polly eats dinner at

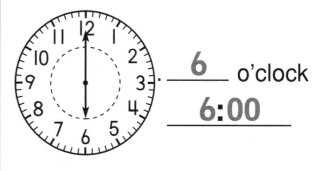

__6__ o'clock

6:00

Mark the correct answer.

5. What time is it?

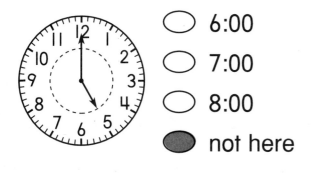

- ○ 6:00
- ○ 7:00
- ○ 8:00
- ● not here

6. What time will it be
   in **two hours**?

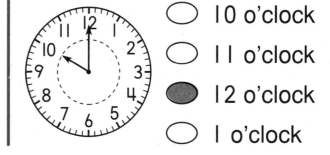

- ○ 10 o'clock
- ○ 11 o'clock
- ● 12 o'clock
- ○ 1 o'clock

Name _____

Hour

Write the time.

1. Mrs. Lee will see the dentist at  .

2:00

2. The station clock says  .

6:00

3. Nora rode her bike at  .

4:00

4. Ted read a book at  .

7:00

Mark the correct answer.

5. Which clock shows the same time?

○ 10:00

● 11:00

○ 12:00

○ 1:00

6. Which clock shows the same time?

3:00

○    ○

○    ●

Time to the Hour

Show the time.
Draw the hour hand and the minute hand.

Check children's drawings.

1. Steve will go
   to the party
   at 2:00.

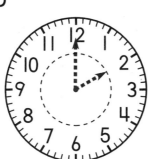

2. Mary will
   get home
   at 5:00.

3. Mrs. Hill will
   go shopping
   at 4:00.

4. Dave goes
   to bed
   at 9:00.

Mark the correct answer.

5. Read the
   clock. What
   time is it?

- ◯ 9 o'clock
- ◯ 10 o'clock
- ◯ 11 o'clock
- ⬤ 12 o'clock

6. Which hand on a clock
   tells the hour?

- ⬤ the short hand
- ◯ the long hand

PROBLEM SOLVING PS101

Half-Hour

Use a . Write the time.

1. Mel's favorite TV show starts at 7:30. It lasts half an hour. What time does it end?

<u>8:00</u>

2. Sunny's dance class starts at 4:30. It lasts one hour. What time does it end?

<u>5:30</u>

Draw the hour and the minute hands.   **Check children's work.**

3. Cathy's music lesson begins at 3:00. It lasts one hour. Show the time it ends.

4. Barb starts to play a game at 10:00. It lasts 30 minutes. Show the time it ends.

Mark the correct answer.

5. What time is it?

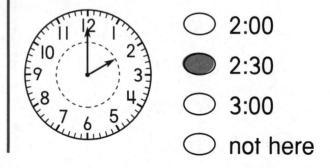

○ 11:30
○ 12:00
● 12:30
○ 1:30

6. What time will it be in half an hour?

○ 2:00
● 2:30
○ 3:00
○ not here

Harcourt Brace School Publishers

Reading Strategy • Use Word Clues

Tina paints a (big) picture
for her dad's birthday.
<u>About how long will it
take to paint the picture?</u>

(more than I minute)

less than I minute

1. Read the problem.
   Draw a line under what you
   want to find out.

2. Circle the word that tells
   about the picture.

3. Estimate how long it will take.
   Circle your estimate.
   Act out the problem to check if needed.

Read the problem. Look for clues
to help you estimate. Circle your
estimate. Then act it out if needed.

4. Tina signs a birthday
   card with her first name.
   About how long will
   it take?

   (less than a minute)

   more than a minute

5. Jim writes a long
   story about his party.
   About how long will
   it take?

   less than a minute

   (more than a minute)

Name _____

Using Nonstandard Units

Draw a picture.
Write how many  long.   **Check children's drawings.**

1. Mark uses 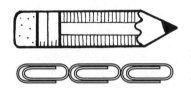 to measure his
   pencil. About how many
   long are 2 pencils?

about ___6___

---

2. Jan uses  to measure a pin.
   About how many  long are
   3 pins?

about ___3___

Mark the correct answer.

3. Dan's desk measures
   29 paper clips long. Bill's
   desk measures 36
   paper clips long. Who
   has the longer desk?

   ○ Dan
   ● Bill

4. The length of Pam's shoe
   is 2 paper clips longer
   than Kim's shoe. Who
   has the longer shoe?

   ● Pam
   ○ Kim

Measuring in Inch Units

Color the inch units.
Write the number of inches.   **Check children's work.**

1. Diane's ribbon is 6 inches long.
   Kara's ribbon is 4 inches shorter.
   How long is Kara's ribbon?

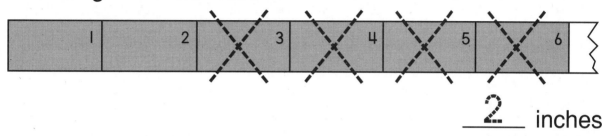

_____2_____ inches

2. Adam's string is 3 inches long.
   Pat's string is 2 inches longer.
   How long is Pat's string?

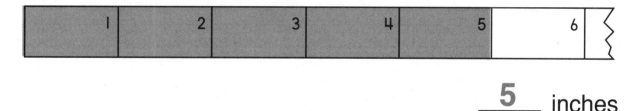

_____5_____ inches

Mark the correct answer.

3. Each link of a chain is
   1 inch. The chain has
   7 links. How long is
   the chain?

   ○ 5 inches
   ○ 6 inches
   ● 7 inches
   ○ 8 inches

4. A toy robot takes 10 equal
   steps to walk a 10-inch
   line. How long is the
   robot's foot?

   ● 1 inch
   ○ 2 inches
   ○ 4 inches
   ○ 5 inches

Using an Inch Ruler

Draw a line to show where to cut.   **Check children's work.**

1. Sarah needs a ribbon 5 inches long.
   Draw a line to show where you would
   cut the ribbon.

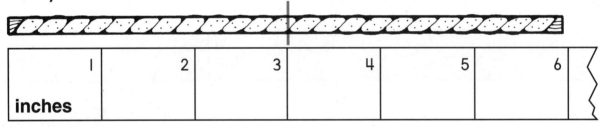

2. Diane needs a piece of yarn 3 inches long.
   Draw a line to show where you would cut
   the yarn.

Mark the correct answer.

3. How long is it?

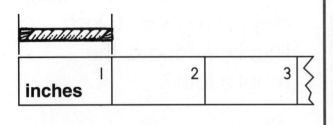

- ○ 2 inches
- ● 1 inch
- ○ 3 inches

4. How long is it?

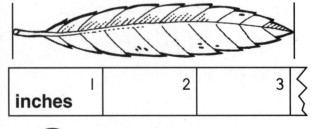

- ○ 1 inch
- ○ 2 inches
- ● 3 inches

Measuring in Centimeter Units

Color the centimeter units.
Write how many.

1. A crayon is 6 centimeters long.
A rubber band is 2 centimeters shorter.
How long is the rubber band?

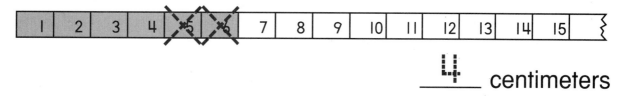

_____4_____ centimeters

2. A pen is 10 centimeters long.
A marker is 2 centimeters longer.
How long is the marker?

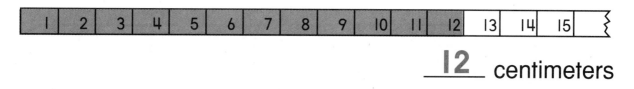

_____12_____ centimeters

Mark the correct answer.

3. How many inches long?

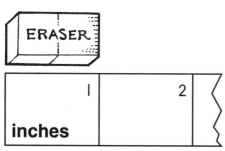

- ● 1 inch
- ○ 2 inches
- ○ 3 inches
- ○ 4 inches

4. How many centimeters long?

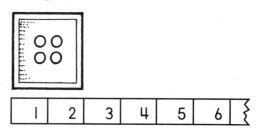

- ● 2 centimeters
- ○ 3 centimeters
- ○ 4 centimeters
- ○ 5 centimeters

Using a Centimeter Ruler

Draw a line to show the centimeters.

1. Mrs. Miller wants to cut
8 centimeters of lace.

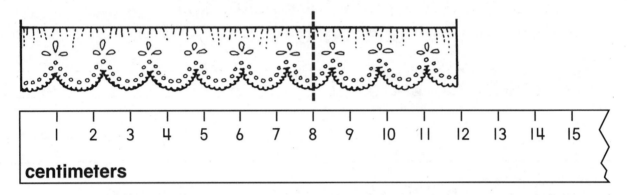

	1	2	3	4	5	6	7	8	9	10	11	12	13	14	15

centimeters

2. Mr. Polt wants to cut
12 centimeters of wood.

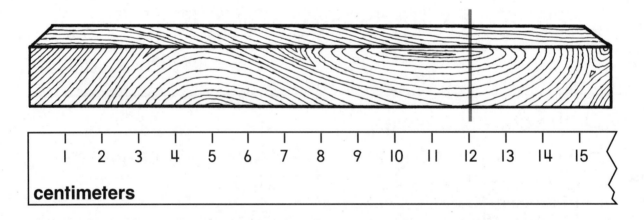

	1	2	3	4	5	6	7	8	9	10	11	12	13	14	15

centimeters

Mark the correct answer.

3. Which object is about
8 centimeters long?

  ○ your math book

  ● a crayon

  ○ a paper clip

4. Which object is about
15 centimeters long?

  ● a paintbrush

  ○ a key

  ○ a safety pin

Using a Balance

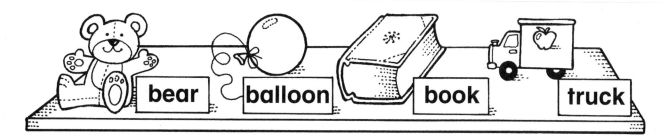

bear   balloon   book   truck

Use the picture. Write the answer.

1. Leslie takes something to share for show-and-tell. She chooses the heaviest object on the shelf.

   What does she choose?   book

2. Rob takes something to share for show-and-tell. He chooses the lightest object on the shelf.

   What does he choose?   balloon

Mark the correct answer.

3. A paper clip is heavier than a pin. A penny is heavier than a paper clip. Which object is heaviest?

   ◯ paper clip
   ◯ pin
   ⬤ penny

4. A shoe is lighter than a book. A tube of paint is lighter than a shoe. Which object is lightest?

   ◯ shoe
   ◯ book
   ⬤ tube of paint

Reading Strategy • Make a Prediction

Making predictions can
help you solve problems.

Marc has a pencil. Rob has
a marker. Which object is heavier?

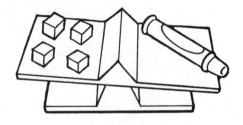

Answers will vary.

1. About how many 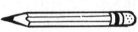 will it take to balance
   the scale? Look at the picture. Make a prediction.

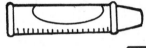

about _____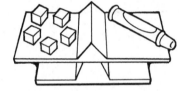

about _____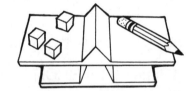

---

2. Measure each object to check.

about **5**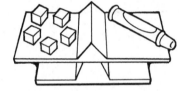

about **3**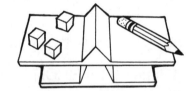

---

3. Which is heavier?

marker

First, make a prediction. **Predictions will vary.**
Then, measure to check.

4. Which is heavier, your
   sock or your shoe?

   shoe

5. Which is heavier,
   1 quarter or 4 nickels?

   4 nickels

Measuring with Cups

Draw a picture.
Solve.

Check children's drawings.

1. A small bowl holds 2 cups of rice.
   A large bowl holds double this
   amount. How many cups does
   the large bowl hold?

   ___4___ cups

2. A carton of juice holds 4 cups.
   Mrs. Jones buys 2 cartons.
   How many cups of juice does
   she buy in all?

   ___8___ cups

3. A blue jug holds 3 cups of milk.
   A red jug holds 2 cups more.
   How many cups does the red
   jug hold?

   ___5___ cups

Mark the correct answer.

4. Which container holds about 1 cup?

Temperature
Hot and Cold

Draw a picture.

Check children's drawings.

1. Jeff plays ball.
   He feels hot.
   Draw something cold for him
   to drink.

2. Sue goes sledding.
   She feels cold.
   Draw something hot for her
   to drink.

3. A jug holds 3 cups of cocoa.
   How many cups will 2
   jugs hold?

   ___6___ cups

Mark the correct answer.

4. Which is hot?

5. Which is cold?

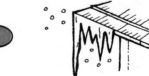

Equal and Unequal Parts of Wholes

Draw lines to solve.   **Check children's drawings.**

1. Kathy and Jane have an orange. They want equal parts. How should they cut the orange?

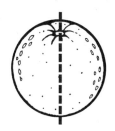

2. Sally and Tom have a sandwich. They want equal parts. How should they cut the sandwich?

3. Four children have a pizza. Each child wants an equal part. How should they cut the pizza?

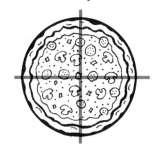

4. Three children have a cake. Each child wants an equal part. How should they cut the cake?

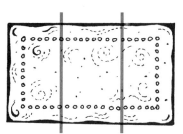

Mark the correct answer.

5. Which figure shows equal parts?

6. Which figure shows equal parts?

PROBLEM SOLVING   PS113

Halves

Draw lines to solve.

Check children's drawings.

1. Lily breaks a cracker to show two equal parts or $\frac{1}{2}$. Show how she could break it.

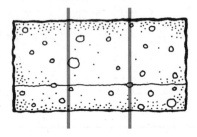

2. Bob and a friend share a pizza. Each gets $\frac{1}{2}$. Show how they could divide it.

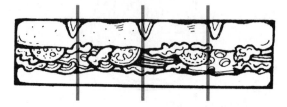

3. Three friends have a large brownie. Each child wants an equal part. How should they cut the brownie?

4. 4 children buy a long sandwich. Each child wants an equal part. How should they cut the sandwich?

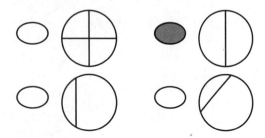

Mark the correct answer.

5. Which figure shows halves?

6. One of two equal parts is _____.

 one half

 one whole

Harcourt Brace School Publishers

Fourths

Color the part to solve.

1. Ruth makes a sandwich. She eats $\frac{1}{4}$ for lunch. Show what part she eats.

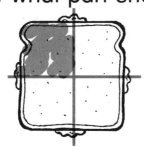

2. Jeff's mom gives him a fruit bar. He eats $\frac{1}{2}$ for a snack. Show what part he eats.

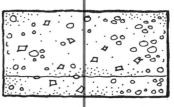

3. Peter picks an apple. He eats $\frac{1}{2}$ for dessert. Show what part he eats.

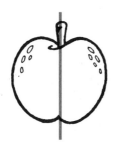

4. Mary bakes a big pizza. She gives $\frac{1}{4}$ to her friend for a treat. Show what part her friend gets.

Mark the correct answer.

5. Which figure shows fourths?

6. Which part is larger?

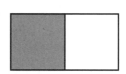

 $\frac{1}{2}$
 $\frac{1}{4}$

Thirds

Color the part to solve.   **Check children's drawings.**

1. Toby buys a large cookie at the snack bar. He eats $\frac{1}{3}$. Show what part he eats.

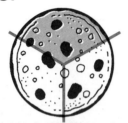

2. Sheila has a big sandwich. She gives $\frac{1}{4}$ to her brother. Show what part her brother gets.

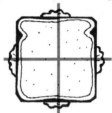

3. Carla has a brownie in her lunchbox. She eats $\frac{1}{3}$ for lunch. Show what part she eats.

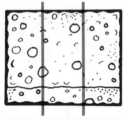

4. Garth buys a pizza. He gives $\frac{1}{2}$ to a friend. Show what part his friend gets.

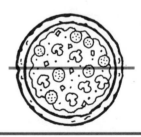

Mark the correct answer.

5. Which figure shows thirds?

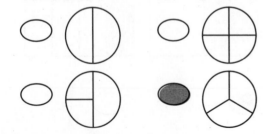

6. A pizza is cut into three equal parts. What do you call one of the equal parts?

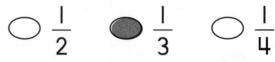

$\bigcirc\ \frac{1}{2}$ $\bullet\ \frac{1}{3}$ $\bigcirc\ \frac{1}{4}$

Harcourt Brace School Publishers

Reading Strategy • Use Visualization

Picturing a problem in your mind can help you
solve the problem.

4 children bake a small cake.
Each child gets an equal part.
How should they cut the cake?

1. Read the problem.
   Picture the four children and the cake.

2. Then picture the cake
   divided into 4 equal parts.

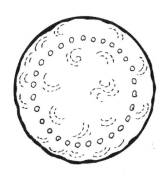

3. Draw lines on the
   cake to show how
   you pictured the cake
   divided into 4 equal parts.

Picture the problem in your mind.
Draw lines to show your picture.   **Check children's drawings.**

4. There are 3 children in all.
   Each one gets an equal
   share of a pie. How
   should they cut the pie?

5. You want to share a pizza
   with 3 other children.
   How should you cut the
   pizza?

Parts of Groups

Draw and color to solve.

Check children's drawings.

1. Regina has 4 toy cars.
   I out of 4 of the toy cars is blue.
   Draw and color the toy cars.

2. Jose has 3 apples.
   I out of 3 of the apples is green.
   Draw and color the apples.

3. There are 4 children and I pizza.
   Each child gets an equal share.
   Draw the pizza.

Mark the correct answer.

4. Which picture shows
   I out of 2 of the plums
   colored?

5. Which picture shows
   I out of 4 of the berries
   colored?

Sort and Classify

Use the table to solve these problems.

Birds at the Feeder	
blue jay	||||
cardinal	||||| |
bluebird	||

1. How many blue jays did Ann see at the feeder?

   __4__ blue jays

2. What kind of bird did Ann see most often at the feeder?

   _____

   - - - - - - - - - - - - - - -

   __cardinal__

3. How many birds with blue feathers did Ann see at the feeder?

   __6__ birds with blue feathers

4. How would Ann show that another blue jay came to the feeder?

   |||||

Mark the correct answer.

5. Which one shows the number 10?

   ○ ||||| ||

   ○ ||||| |||

   ○ ||||| ||||

   ● ||||| |||||

6. How are the birds in Ann's table sorted?

   ○ by age

   ○ by color

   ● by kind

   ○ by size

Certain or Impossible

Use the picture.
Color the answer
to each question.

1. Sam is hungry. What can he eat from his bag?

2. Sam wants to buy a snack. What can he use from his bag?

3. Sam wants to color a picture. What can he use from his bag?

4. Sam needs to write. What can he use from his bag?

Mark the correct answer.

5. Which of these things is Sam **certain** to find in his bag?

6. Which of these things is it **impossible** for Sam to find in his bag?

Most Likely

Draw a picture.
Write the answer.

1. A bag has 6 apples and 2 oranges. Jane closes her eyes and takes out a fruit. Which fruit is she most likely to get?

   apple

2. A bag has green grapes and red grapes. Holly wants purple grapes. Can Holly get purple grapes from the bag?

   _No_

Mark the correct answer.

3. What is your prediction for choosing a 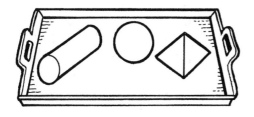?

   ● impossible

   ○ certain

4. What is your prediction for choosing a △?

   ○ impossible

   ● certain

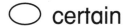

Reading Strategy • Use Graphic Aids

You can use a table to help
you solve the problem.

Nicky has a spinner.
She predicts the spinner will stop
on blue more times than red.

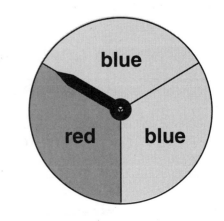

1. Which color do you predict the
   spinner will stop on more often?

   _____

   _____
         Answers will vary.

2. When Nicky spins, she lands on
   blue 7 times and on red 3 times.
   Make a tally mark
   on the table for
   each spin.
   Write the totals.
   Check Nicky's
   prediction.

	Tally Marks	Total
blue	||||| ||	7
red	|||	3

Solve.

3. Tonia has a bag with 4 red cubes and 8
   white cubes. She predicts she will get a
   white cube more often than a red cube.
   When she tries, she gets red 2 times and
   white 8 times. Is her prediction a good one?

    yes   no

Harcourt Brace School Publishers

Picture Graphs

Favorite Pets					
dog	🐶	🐶	🐶	🐶	
bird	🐦				
fish	🐟	🐟	🐟		
cat	🐱	🐱	🐱	🐱	🐱

Use the graph to answer the questions.

1. How many children in the class choose fish as their favorite pet?

___3___ children

2. Write a number sentence that tells how many children like fish and birds best.

__3__ (+) __1__ = __4__

3. 4 more children choose birds as their favorite pet. How many children like birds now?

___5___ children

4. The children who like cats and dogs best work together to make a picture. How many children work on the picture?

___9___ children

5. Which pet do most children like best?

○ dog

● cat

○ fish

○ bird

6. How many children in the class like mice best?

○ 5 children

○ 2 children

○ 1 child

● not here

Reading Strategy • Compare and Contrast

Ramon makes a graph to find out
which fruit his friends like best.

Favorite Fruits

blueberries						
peaches						
apples						
bananas						

0 1 2 3 4 5 6

1. Look at the graph.
Compare the colored squares in each row.
Which fruit has the most colored squares?

_____

peaches

2. Which kind of fruit do most children like best?

_____

peaches

3. Which kind of fruit do the
fewest children like?

bananas

4. Do more children like
apples or blueberries?

apples

Vertical Bar Graphs

1. Color the graph to match the tally table.

Favorite Colors		Total
Green	||	2
Red	|||	3
Blue	|	1

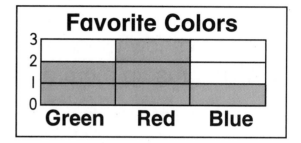

Favorite Colors

3
2
1
0
Green   Red   Blue

2. Which color do most children like best?

___red___

3. How many more children like green better than blue?

___1___ child

4. How many children like red and blue?

___4___ children

5. Which color did the fewest children like?

___blue___

Mark the correct answer.

6. Which number sentence shows how many more children liked red than blue?

● $3 - 1 = 2$

○ $3 - 2 = 1$

○ $2 + 1 = 3$

7. How many children in all named their favorite colors?

○ 4 children

○ 5 children

● 6 children

○ not here

Reading Strategy • Use Graphic Aids

Nancy flips a coin 10 times.
She shows what happens in
a tally table. Then she makes
a graph to show which side of
the coin turns up more often.

Nancy's Coin Flips	
heads	IIII
tails	⊬⊬ I

1. Look at the tally table.
   Count the tally marks and write the totals.

Nancy's Coin Flips		Total
heads	IIII	4
tails	⊬⊬ I	6

2. Color the graph to match the tally table.

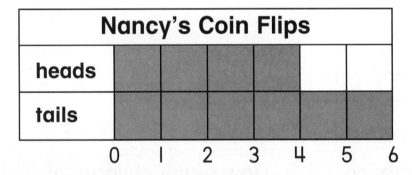

Mark the correct answer.

3. Which side of the coin
   turned up more often?

   ⬤ tails

   ◯ heads

4. Do you think that tails will
   always turn up more
   often than heads?

   ◯ yes

   ⬤ no

Harcourt Brace School Publishers

Doubles Plus One

Draw a picture. Write the sums.   **Check children's drawings.**

1. There are 6 bluebirds sitting on a fence. Then 7 more join them. How many bluebirds are there in all?

   __6__ + __7__ = __13__ bluebirds

2. There are 8 crows eating corn. Then 9 more crows come. How many crows are there in all?

   __8__ + __9__ = __17__ crows

3. Tim draws 5 birds. Sue draws 6 more birds. How many birds do the children draw in all?

   __5__ + __6__ = __11__ birds

Mark the correct answer.

4. Which doubles fact can help you solve

   $4 + 3 =$ ____?

   ● $4 + 4$ ○ $5 + 5$
   ○ $7 + 7$ ○ $8 + 8$

5. Which doubles fact can help you solve

   $8 + 9 =$ ____?

   ○ $5 + 5$ ○ $6 + 6$
   ○ $7 + 7$ ● $8 + 8$

Doubles Minus One

Check children's drawings.

Draw a picture. Write the sums.

1. Mrs. Park has 7 stamps. She buys 6 more. How many stamps does she have in all?

   __7__ + __6__ = _13_ stamps

2. Leroy mails 5 letters in the morning and 7 letters in the afternoon. How many letters does he mail in all?

   __5__ + __7__ = _12_ letters

3. Mr. Jones buys 9 baseball stamps and 8 basketball stamps. How many stamps does he buy in all?

   __9__ + __8__ = _17_ stamps

Mark the correct answer.

4. Which doubles fact can help you solve

   $5 + 4 =$ ____?

   ○ 3 + 3   ● 5 + 5
   ○ 7 + 7   ○ 8 + 8

5. Which doubles fact can help you solve

   $8 + 7 =$ ____?

   ○ 6 + 6   ○ 9 + 9
   ● 8 + 8   ○ not here

Harcourt Brace School Publishers

Doubles Patterns

Draw a picture. Solve.

Check children's drawings.

1. A basketball team has 5 players.
   Two teams meet for a game, but
   I player does not come.
   How many players are there?

   __*9*__ players

2. Two baseball teams play a game.
   Each team has 8 players. One
   team has I extra player. How
   many players are there in all?

   __17__ players

3. A soccer team has 7 players.
   Two teams play a game.
   I player has to go home.
   How many players are left?

   __13__ players

Mark the correct answer.

4. Which is a doubles-minus-
   one fact?

   ○ $9 + 9 = 18$

   ○ $9 + 7 = 16$

   ● $9 + 8 = 17$

   ○ $9 + 10 = 19$

5. Which is a doubles-plus-
   one fact?

   ○ $4 + 4 = 8$

   ○ $5 + 5 = 10$

   ○ $5 + 4 = 9$

   ● $5 + 6 = 11$

Harcourt Brace School Publishers

PROBLEM SOLVING   PS129

Doubles Fact Families

Draw a picture.
Add or subtract to solve.

Check children's drawings.

1. Steve had 18 marbles.
   He lost 9. How many marbles
   does he have left?

   ___9___ marbles

2. Cindy has 8 rocks. She finds 7
   more. How many rocks does
   she have in all?

   ___15___ rocks

3. Joe finds 8 shells. Larry finds two
   times as many as Joe. How
   many shells does Larry find?

   ___16___ shells

Mark the correct answer.

4. Which group of three
   numbers can be used
   to make a doubles
   fact family?

   ● 4, 4, 8   ○ 4, 6, 8
   ○ 2, 4, 6   ○ 4, 8, 12

5. Which number sentence
   belongs in a doubles
   fact family?

   ○ $6 + 8 = 14$
   ● $7 + 7 = 14$
   ○ $14 - 6 = 8$
   ○ $14 - 8 = 6$

Harcourt Brace School Publishers

Reading Strategy • Use Word Clues

Sal has 3 pencil toppers. Nora has **two times** as many. How many pencil toppers do they have **in all**?	Ken had 16 animal erasers. He **gave some away**. He has 8 left. How many did he **give away**?

1. Read each problem. Look for word clues. Think. How do these words help me decide if I should add or subtract?

---

2. Use counters to solve each problem. Draw them. Write the answers.

_9_ pencil toppers

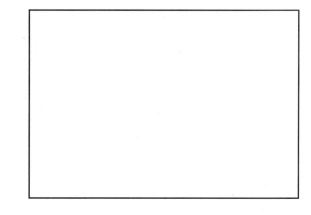

_8_ animal erasers

Solve.

3. Jeff has 5 yo-yos. Noah has 1 more than Jeff. How many yo-yos do the boys have in all?

_11_ yo-yos

4. Margie had 17 stickers. She gave some away. She has 9 left. How many did she give away?

_8_ stickers

Make a 10

Use the make-a-ten strategy to add.
Draw a picture.

Check children's drawings.

1. Ricky has 8 baseball cards.
   He buys 2 more. How many
   cards does he have in all?

   __10__ cards

2. Sally has 7 jacks.
   She finds 5 more. How many
   jacks does she have in all?

   __12__ jacks

3. Carl has 9 stickers. His sister
   gives him 4 more. How many
   stickers does he have in all?

   __13__ stickers

Mark the correct answer.

4. Which fact belongs with
   these facts?

   $2 + 8$ $6 + 4$ $5 + 5$

   ○ $4 + 5$ ○ $8 + 1$

   ● $7 + 3$ ○ $9 + 3$

5. Which shows how to use
   the make-a-ten strategy
   to add $6 + 5$?

   ● $10 + 1 = 11$

   ○ $5 + 6 = 11$

   ○ $5 + 5 + 1 = 11$

Adding Three Numbers

Write a number sentence.
Solve.

1. Liz has 8 red beads, 8 green beads, and 1 yellow bead. How many beads does she have in all?

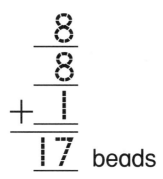

$$\begin{array}{r} 8 \\ 8 \\ +\ 1 \\ \hline 17 \end{array}$$ beads

2. Jim has 1 blue marble, 2 green marbles, and 9 red marbles. How many marbles does he have in all?

$$\begin{array}{r} 1 \\ 2 \\ +\ 9 \\ \hline 12 \end{array}$$ marbles

Mark the correct answer.

3. Which strategy would you use to find this sum?

$5 + 6 + 4 =$ _____

○ doubles
⬤ make a ten
○ count on 3

4. Which problem can you use a doubles strategy to solve?

⬤ $6 + 6 + 5$
○ $6 + 5 + 4$
○ $6 + 7 + 8$

Sums and Differences to 14

Draw a picture.
Solve.

Check children's drawings.

1. Randy has 8 puppets. He gets 3 more for his birthday. How many puppets does he have in all?

   _**11**_ puppets

2. Randy has 11 puppets. He leaves 3 at school. How many puppets does he have now?

   _**8**_ puppets

3. Jane has 7 yellow trolls, 2 red trolls, and 5 green trolls. How many trolls does she have in all?

   _**14**_ trolls

Mark the correct answer.

4. Which subtraction sentence is related to this addition sentence?

   $$7 + 9 = 16$$

   ○ $16 - 8 = 8$
   ● $16 - 7 = 9$
   ○ $9 - 7 = 2$

5. Which three numbers could be used to make related addition and subtraction sentences?

   ● 5, 8, 13
   ○ 6, 9, 12
   ○ 7, 5, 11

Harcourt Brace School Publishers

Sums and Differences to 18

Draw a picture to solve.

Check children's drawings.

1. There are 8 children at the playground. 9 more children come. How many children are at the playground?

   ___17___ children

2. 3 children ride bikes. 5 children play on the slide. 7 children swing. How many children are playing in all?

   ___15___ children

3. There are 17 children at the playground. 8 go home for lunch. How many children are left?

   ___9___ children

Mark the correct answer.

4. Which strategy could you use to solve this problem?

   $$6 + 7 = \underline{}$$

   ○ doubles

   ○ make-a-ten

   ● doubles plus one

5. What subtraction fact can you figure out by knowing

   $$8 + 7 = 15?$$

   ○ $15 - 10 = 5$

   ● $15 - 7 = 8$

   ○ $15 - 9 = 6$

Counting Equal Groups

Draw a picture.
Write how many in all.

Check children's drawings.

1. Mrs. Jones has 4 window boxes.
She plants 3 flowers in each box.
How many flowers in all?

___12___ flowers

2. Mr. Hill has 3 pots.
He plants 3 seeds in each pot.
How many seeds in all?

___9___ seeds

3. Ms. Green has 5 plants.
Each plant has 2 flowers.
How many flowers in all?

___10___ flowers

Mark the correct answer.

4. Which picture shows
2 groups of 3?

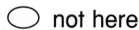 not here

5. Which addition sentence
shows 3 groups of 4?

○ $4 + 3 = 7$

● $4 + 4 + 4 = 12$

○ $3 + 3 + 3 + 3 = 12$

○ not here

Harcourt Brace School Publishers

How Many in Each Group?

Draw a picture.
Solve.

1. Larry feeds 9 bananas to 3 monkeys. Each monkey gets the same number of bananas. How many bananas does each monkey get?

   __3__ bananas

2. Pam feeds 12 fish to 4 seals. Each seal gets the same number of fish. How many fish does each seal get?

   __3__ fish

3. Larry feeds 3 goats. He gives them 4 cups of food each. How many cups of food does Larry need?

   __12__ cups

Mark the correct answer.

4. Which is greater?

   ● 3 groups of 4

   ○ 3 groups of 3

5. Which has more groups?

   ○ 2 groups of 5

   ● 5 groups of 2

How Many Groups?

Draw a picture.
Solve.

Check children's drawings.

1. Art has 12 rolls. He puts 6 rolls in each basket. How many baskets does he use?

   ___2___ baskets

2. Noah has 3 friends. He gives them each 3 crackers. How many crackers does he need?

   ___9___ crackers

3. Trudy has 8 cookies. She puts 2 cookies on each plate. How many plates does she use?

   ___4___ plates

Mark the correct answer.

4. Which picture shows 4 groups of 3?

   ○ not here

5. Which addition sentence shows adding 3 groups of 5?

   ○ $3 + 5 = 8$

   ● $5 + 5 + 5 = 15$

   ○ $3 + 3 + 3 + 3 + 3 = 15$

Harcourt Brace School Publishers

Reading Strategy • Use Word Clues

Using word clues can help
you solve problems.

There are **3 bowls** of fruit salad. **Each bowl has 2** red grapes. How many red grapes **in all**?	There are **12 cherries** and **3 bowls**. How many cherries **in each bowl**?	There are **15** green **grapes**. Anna put **5 grapes in each bowl**. **How many bowls** did she use?

1. Read the problem. Look for word clues.

2. Use these word clues to help you.
   Draw a picture. Then solve.

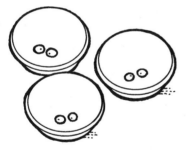

6 red grapes
in all

4 cherries in
each bowl

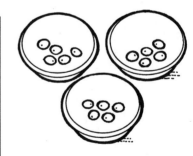

3 bowls

Solve.

3. There are 9 crackers.
   Each child gets 3.
   How many children
   get crackers?

   ___3___ children

4. There are 3 children.
   Each child gets 2 cups of
   juice. How many cups of
   juice in all?

   ___6___ cups of juice

Adding and Subtracting Tens

Draw a picture.
Find the sum or difference.

Check children's drawings.

1. Mr. Ellis has 50 cows. He buys 10 more. How many cows does he have in all?

   __60__ cows

2. Mrs. Ellis has 90 blocks of cheese. She sells 20 blocks. How many blocks of cheese does she have left?

   __70__ blocks of cheese

3. Kathy milks 30 cows. Bill milks 20 cows. How many more cows does Kathy milk?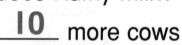

   __10__ more cows

Mark the correct answer.

4. Find the sum.

$$\begin{array}{r} 30 \\ +20 \\ \hline \end{array}$$

   ⚪ 5   ⚪ 30

   ⬤ 50   ⚪ 55

5. Find the difference.

$$\begin{array}{r} 70 \\ -60 \\ \hline \end{array}$$

   ⚪ 70   ⚪ 60

   ⚪ 50   ⬤ 10

Harcourt Brace School Publishers

Adding Tens and Ones

Draw a picture.
Find the sum.

Check children's drawings.

1. Ned spends 57¢ on a notebook and 11¢ on a pencil. How much does Ned spend in all?

   68 ¢

2. Phil has 50¢ in his pocket. He earns 20¢ more. How much money does Phil have?

   70 ¢

3. Karen wants to buy 2 cards. Each card costs 30¢. How much money does Karen need?

   60 ¢

Mark the correct answer.

4. Add.

tens	ones
5	2
+2	6

   ○ 87
   ○ 77
   ● 78

5. When adding tens and ones, begin by adding the _____ first.

   ● ones
   ○ tens

Subtracting Tens and Ones

Draw a picture.
Find the sum or difference.

1. Mrs. Jones makes 46 cookies.
She gives 12 cookies to Tim
and his friends. How many
cookies are left?

   __34__ cookies

2. The Cub Scouts make 20 tacos.
They eat 10 for lunch.
How many tacos are left?

   __10__ tacos

3. Mary makes 24 trail mix bars.
Paula makes 24 also.
How many trail mix bars do
they make?

   __48__ trail mix bars

Mark the correct answer.

4. Subtract.

tens	ones
4	5
−3	1

   ◯ 76

   ◯ 41

   ⬤ 14

5. When subtracting
tens and ones, begin
by subtracting the
_____ first.

   ⬤ ones   ◯ tens

Reading Strategy • Use Word Clues

Using word clues can help you decide
if an answer to a problem makes sense.

José has **61 shells**.
He finds **12 more**.
How many shells does he have in all?

7 shells   73 shells   730 shells

1. Read the problem. Look for word
   and number clues to help you find
   the answer that makes sense.

2. Decide if the answer is likely to be
   in the ones, tens, or hundreds.

3. Choose the answer that makes sense.

_73_ shells

Mark the answer that makes sense.

4. Mr. Williams has 47
   stamps. He uses 42.
   How many stamps does
   he have left?

   ● 5 stamps

   ○ 89 stamps

   ○ 500 stamps

5. Alice has 53 marbles.
   She buys 35 more.
   How many marbles she
   have now?

   ● 88 marbles

   ○ 880 marbles

   ○ 22 marbles